D1258009

THIRD EDITION

The Global Positioning System and ArcGIS

THIRD EDITION

The Global Positioning System and ArcGIS

Michael Kennedy
University of Kentucky

Featuring hardware and GPS software from
Trimble Navigation, Limited, and GIS software
from Environmental Systems Research Institute (ESRI)

CRC Press
Taylor & Francis Group
Boca Raton London New York

CRC Press is an imprint of the
Taylor & Francis Group, an **informa** business

CRC Press
Taylor & Francis Group
6000 Broken Sound Parkway NW, Suite 300
Boca Raton, FL 33487-2742

© 2010 by Taylor and Francis Group, LLC
CRC Press is an imprint of Taylor & Francis Group, an Informa business

No claim to original U.S. Government works

Printed in the United States of America on acid-free paper
10 9 8 7 6 5 4 3 2 1

International Standard Book Number: 978-1-4200-8799-4 (Hardback)

Library of Congress Cataloging-in-Publication Data

Kennedy, Michael, 1939-
    The global positioning system and ArcGIS / Michael Kennedy. -- 3rd ed.
        p. cm.
    Includes bibliographical references and index.
    ISBN 978-1-4200-8799-4 (hardcover : alk. paper)
    1. Global Positioning System. 2. Geographic information systems. I. Title.

G109.5.K46 2010
623.89'3--dc22
                                                                            2009018854

Visit the Taylor & Francis Web site at
http://www.taylorandfrancis.com

and the CRC Press Web site at
http://www.crcpress.com

for Alexander Kennedy

Contents

Foreword to the First Edition

Michael Kennedy's latest book brings together Geographic Information System (GIS) technology and Global Positioning System (GPS) technology with the aim of teaching how to couple them to effectively capture GPS data in the field and channel it to a GIS.

We at the Environmental Systems Research Institute (ESRI) were especially pleased that he chose to use ESRI's GIS software (ArcInfo and ArcView) in writing his book, and we were happy to be able to provide him with some support in his efforts.

After 15 or 20 years in which very few textbooks were written about GIS and related technologies, there is now a veritable flood of new GIS books coming into print. Why is this one especially valuable?

First, because it couples *GPS/GIS* in an especially intimate way. Michael's intention in writing was to make it possible for readers working alone or for students in a formal course to learn how to use *GPS/GIS* "hands on," to walk away from this textbook ready to go into the field and start using Trimble Navigation's GeoExplorer and ESRI's ArcInfo and ArcView software to collect GPS field data and enter it into a GIS for immediate use.

Besides providing step-by-step instructions on how to do this, he provides appropriate background information in the form of theoretical discussions of the two technologies and examples of their use. He writes in an easy style, explaining the needed technical and scientific principles as he goes, and assuming little in the way of necessary prior instruction.

Instructors will find the text especially useful because Michael provides the kinds of detailed procedures and hints that help to make lab work "bulletproof" even for the inexperienced student. An accompanying CD-ROM has data that are likely to be useful in various ways to teacher and student alike.

Second, this book is important because *GPS/GIS* is such an extremely important technology! It is no exaggeration to say that *GPS/GIS* is *revolutionizing* aspects of many fields, including surveying (slashing the costs of many kinds of survey efforts and bringing surveying to parts of the world where surveys are nonexistent, highly inaccurate, or long since outdated), the natural resource fields (providing rapid and far more accurate collection of field natural resource data of many kinds), and municipal planning (providing for the updating of all kinds of records based on accurate field checking), to name only a few. GPS is making practical the kinds of data collection that were simply out of the question only a few years ago because the necessary skilled teams of field personnel were unavailable and the costs of accurate field data collection were beyond the means of virtually all organizations that needed these kinds of data. *GPS/GIS* is changing all that. Use of the kinds of methods taught in *The Global Positioning System and GIS* is spreading very rapidly; *GPS/GIS* use will become commonplace throughout dozens of fields in just the next few years as costs of hardware and software continue to fall and books like this one increase the number of persons familiar with these two coupled technologies.

The impact of this revolution in data gathering will, I believe, have profound effects on the way in which we view the earth, on ways in which we exercise our stewardship of its resources for those who come after us, and on the professional practice of an extraordinary range of disciplines (engineering, oceanography, geology, urban planning, archaeology, agriculture, range management, environmental protection, and many, many others). I look forward to the time when *tens of millions of people* will make use of *GPS/GIS* technology every day, for thousands of purposes.

ESRI's aim as a company has always been to provide reliable and powerful GIS and related technologies to our clients and users and to help them use these technologies to make their work more effective and successful. By doing so, we hope to help make a difference in the world.

The ESRI authors' program was created to further those same goals. Michael Kennedy's *The Global Positioning System and GIS* is an extremely important part of that program and will, I believe, assist many persons to acquire and effectively use *GPS/GIS* in their work. In writing this book he has performed an important service, not just to his readers and the users of this textbook but also to those whose lives will be improved because of the use of *GPS/GIS* technologies in the years ahead.

Jack Dangermond
President
ESRI, Inc.

Foreword to the Second Edition

Michael Kennedy's timely revision of *The Global Positioning System and GIS* has occurred during a period of significant change in both GPS and GIS technologies. First, with GPS, the U.S. Department of Defense has removed Selective Availability, the major source of error in GPS position; second, "location," as an attribute, is becoming integrated into everyday business process with the advent of Location-Based Services.

The first edition of his book contains several predictions about the future use of GPS. Many of these have come true. Accuracy has improved, both gradually by improvements in the GPS receiver hardware and significantly by the removal of Selective Availability. Air navigation is seeing transformation with the FAA's Wide Area Augmentation System (WAAS) being close to operational status. GPS is being used as a time base in most time-critical applications such as the Internet and cellular telephone networks, and we are seeing the evolution of applications that a few years ago had not been considered—for example, the use of GPS to provide automatic guidance in agricultural applications, which in turn raises the productivity of scarce land resources.

Michael Kennedy also predicted that GPS in combination with other systems and technologies would provide positional information in places of poor or limited reception. This has certainly happened, but it is the combination of reliable and accurate positioning, broadband wireless network technology, and the Internet that is driving the need for a whole variety of additional types of spatial data that forms the underlying structure of Location-Based Services. GIS is moving from a supporting, backroom analysis tool to mission-critical status as part of a 24/7 real-time management tool. Combining the changes that have occurred in GPS to those that are occurring today in GIS with the addition of wireless networks and the Internet creates a technology market space that is as exciting as the initial GPS/GIS market was when the first edition was written.

Trimble's new GeoExplorer 3 has been used for the updated examples in the text. This product typifies the improvements that have been made in GPS/GIS technology, not only the improvements in the basic GPS technology but also, probably more importantly, the improvements in the "ease of use" functions such as the user interface for feature and attribute collection. Functionality for the updating of spatial databases in addition to the primary function of data collection are new in the GeoExplorer 3, highlighting the importance of maintaining existing data, as the analysis based on the spatial data is only as accurate as the underlying database.

As the use of GPS technology becomes more a part of our daily lives, from timing the Internet to proving position within mobile phones and location information for a range of location-based services, the need for professionals skilled at collecting and maintaining spatial databases can only increase. Michael Kennedy's second edition continues the theme of a clear step-by-step instruction on how to combine ESRI's software with Trimble's GeoExplorer to collect field data and transfer this to a GIS while understanding the theory behind the technologies.

Trimble is pleased to again be associated with Michael Kennedy's *The Global Positioning System and GIS*, and by providing this revision he continues to provide the basis by which students of the text can gain a fundamental and practical understanding of how GPS/GIS can be applied in a vast range of applications.

Alan Townsend
Vice President Mapping and GIS Division
Trimble Navigation, Ltd.

Preface for the Instructor, Third Edition*

THE THIRD EDITION—WHAT'S DIFFERENT

The major difference between this text and the first two editions (1996 and 2002) is that I have made the text applicable to an array of receivers and software. Specifically included are the Trimble Juno series (running TerraSync) and the Magellan MobileMapper (running ESRI ArcPad). The text has been constructed so that regardless of the data capture mechanism, the reader or student will be able to collect GPS data and install it in ArcGIS. The text also has many examples of GPS/GIS data. The student gets to see GPS fixes superimposed on digital orthophotoquads, on topographic quadrangles digital raster graphic form, on soils and geologic coverages, on TIGER files, and so on. The primary lesson of the book—that the issues of datum, coordinate system, projection, and units are vital to correctly integrating GPS and GIS data—is reinforced.

There is more emphasis on ArcGIS, with the assumption that it will be the usual target of GPS conversion activities.

The entire text has been combed through, carefully tested, and re-tested. Figures and photos have been added to make the exercises flow more smoothly.

PURPOSE AND AUDIENCE

The purpose of this textbook/workbook and the accompanying CD-ROM is to provide a short, intermediate, or full-term course in using the Global Positioning System (GPS) as a method of data input to ESRI's ArcGIS. The short course may either stand alone or be a 2- to 4-week segment in a general course in GIS. The text may be used in a formal course with several students or as an individual self-teaching guide. There is the assumption that the students have at least a passing familiarity with ArcGIS and with the most basic geographical concepts, such as latitude and longitude—but the book can serve as a gentle introduction to GPS even for those who do not intend to use the data in a GIS.

A LOOK AT THE CONTENTS

First, a note. If you have used an earlier edition of this text you will notice a number of enhancements and changes present in the new edition. I will point out the significant ones at the end of the discussion of each chapter below.

---

* Material for the student begins with the Introduction.

Chapter 1, "Basic Concepts," is an introduction to GPS as a system. The field/lab work involves data collection with pencil and paper. This works best if the data collection point involves a surveyed-in monument, but it's not necessary. *New*: Both Trimble TerraSync and ESRI ArcPad are used to collect data. If you are using some other data collection software the text is still useful, but will require more work on the student's part—specifically developing a Fast Facts File for information relating to your specific receiver and software. Three different text box outlines are used to identify the hardware or software being described.

Chapter 2, "Automated Data Collection," looks further into how GPS works. Project work includes taking GPS data and storing it in files, first in a receiver, then in a PC. You might want to have the students take data in the same place in Chapter 2 as in Chapter 1. *New*: In addition to TerraSync and Pathfinder Office we look at ArcPad, running on a Magellan MobileMapper. Differential Correction with SBAS (Satellite-Based Augmentation Systems such as WAAS, EGNOS, or MSAS) are discussed and such corrections applied to both TerraSync and ArcPad data collection.

Chapter 3, "Examining GPS Data," initially answers some questions to complete the discussion of the theory of GPS, and then gives the reader experience with software for processing and displaying collected GPS data. GPS data sets are superimposed on digital orthophotoquads (DOQs). *New*: When possible, datasets are examined using both Pathfinder Office and ArcMap.

Chapter 4, "Differential Correction, DOQs, and ESRI Data," discusses the issue of accuracy and techniques for obtaining it. Practice first takes place with "canned" data, then with the user's own. *New*: Looking at SBAS (WAAS) correction in light of "traditional" differential correction.

Chapter 5, now titled "Integrating GPS Data with GIS Data," has the reader exporting GPS for ArcGIS. *New*: Using ArcMap, the student sees some very basic GPS data in the context of many other types of GIS data. Data sets include raster DRGs, DOQs, DEMs, and various vector data sets such as land use, soil type, geology, political boundaries, and TIGER files.

Chapter 6, "Attributes and Positions," discusses the highly important matters of collecting feature attribute data as position data is obtained, and installing it in a GIS. *New*: Emphasis is on ArcMap as well as Pathfinder Office.

Chapter 7, "GPS Mission Planning," assures that one can take data at the best times.

Chapter 8, "The Present and the Future," is a warning about privacy concerns with GPS-enabled devices, a brief discussion of the issues and uses of GPS today and some guesses about what is to come.

There are two Appendices. Appendix A gives additional sources for GPS information—mostly Internet pointers. I do not include a glossary of GPS terms, but point to some on the Internet. Appendix B is a form you may use to challenge students with data sets included on the CD-ROM. Those GPS datasets may be used for exercises, lab practicals, or exams.

THE LEARNING APPROACH: BOTH THEORETICAL AND HANDS-ON

The approach I use is to divide each chapter into two modules:

- Overview
- Step-by-Step

This division comes from my belief that learning a technical subject such as GPS involves two functions: education and training. The student must understand some of the theory and the language of the subject he or she is undertaking, but the ability to perform tasks using the technology is likewise important, and such hands-on experience provides new insights into the subject. Many other books, on other technical subjects, attempt to perform both of these functions, but mix the functions together. It is my view—particularly in the case of GPS, which involves several complex systems, field work, and learning about hardware and software—that the two functions serve the learner best if they are distinctly separated. The theory is presented in the Overview module. It is laid out in a hierarchical, simple-to-complex, way (from the top, down). The training portion takes place in a linear (step-by-step) form: "do this; now do this."

There is heavy emphasis on getting the vital parameters—data, projection, coordinate system, units—correct. No conscientious student should be able to leave a course based on this book and commit the all-too-easy sin of generating incorrect GIS information based on incorrectly converted GPS data.

HARDWARE AND SOFTWARE

Teaching the U.S. Global Positioning System (NAVSTAR) using GIS with a "hands-on" approach involves using particular hardware and software. I decided it was not possible to write a satisfactory text that was general with respect to the wide variety of products on the market. For hardware I therefore selected the two of the most popular and capable products available: the Trimble Navigation Juno series (ST, SA, and SB) and the Magellan MobileMapper GPS receivers. For GIS software I emphasize ArcGIS Desktop from Environmental Systems Research Institute (ESRI). If you choose to use other products—and there are several fine ones in both GPS and GIS—you may still use the text, but you will have to be somewhat creative in applying the Step-by-Step sections.

AN INSTRUCTOR'S GUIDE: ON THE CD-ROM

An instructor who undertakes to teach a GPS/GIS course for the first time may face a daunting task, as I know only too well. In addition to many of the problems that face those teaching combination–lecture-and-lab courses, there are logistical problems created by the (assumed) scarcity of equipment, managing teams of students, the need to electrically charge the receiver batteries, the outdoors nature of some of the lab work, and other factors. I have made an attempt to keep other teachers from encountering the difficulties I found, and to anticipate some others.

Please take a bit of time to investigate the CD-ROM. No special installation of the CD-ROM is required. All the files and folders may simply be copied to the location you desire using standard Windows procedures. To use the CD-ROM start by reading the file named **"READTHIS.1ST"**. It is available as both a simple text file that may be opened in any text editor (e.g., WORDPAD) and as a MS DOC file. This file will indicate how to access the "Instructor's Guide" and how to copy (from the CD-ROM to the root directory of any drive you wish on your machine) the principle folders of data required for the books projects: These folders are GPS2GIS (containing Pathfinder Office data) and IGPSwArcGIS (consisting of ESRI shapefiles).

The CD-ROM also contains other helpful sections. For example,

- Digital copies of the paper forms in the text (such as latitude-longitude computations, equipment checkout and setup, and base station information) are provided so you may modify them for your situation and print them out for distribution to your students
- Outlines for courses of varying lengths and hands-on-involvement levels
- How to obtain a solution manual for the text containing answers to questions and exercises posed in the text

DEMONSTRATION DATA: ON THE CD-ROM

The use of the text requires that some of the demonstration data on the CD-ROM be copied to a hard drive on your machine. The text is quite flexible with respect to use of data supplied on the CD-ROM in conjunction with the data the students, usually working in pairs, collect on their own. Most assignments begin with manipulating the canned data, to prevent surprises, followed by students using home-grown data, if time and interest permit.

In addition to the data for the exercises in the Step-by-Step sections of the book, the CD-ROM contains a great number of files of GPS data from North America and, to a lesser extent, from other parts of the world. Some of these are within additional exercises, detailed on the CD-ROM, that may be used for student practice or student testing.

The Step-by-Step procedures have been tested, and tested again, so the projects and exercises should work as indicated—subject, of course, to new releases of software, firmware, and hardware.

Acknowledgments

FIRST EDITION

The project that led to the first edition of this text began in a conversation with Jack Dangermond of ESRI in May of 1993. I proposed to quickly put together a short introduction to GPS for GIS users. Since that time, I have learned a great deal about both GPS and "quickly putting together a short introduction." Several software releases later, several introductions of GPS hardware later, and many occasions on which I learned that there was a lot more to this subject than I imagined, it is finished. I could not have done it without the following people:

My daughter, Heather Kennedy, who gave great help and support along the way and, particularly, who did the painstaking and intense work of final proofing and testing.

My son, Evan Kennedy, who added to the collection of GPS tracks presented in the text by taking a GPS receiver across the United States by automobile, and whose enthusiasm for GPS encouraged me to complete the book.

Allan Hetzel, who took on the job of finalizing the computer files. He coped with corrections from several reviewers and coordinated the final marathon production session.

Dick Gilbreath, who, with the help and forbearance of Donna Gilbreath, spent many hours at inconvenient times producing most of the figures, and who insisted on getting the smallest details right.

Yu Luo and Pricilla Gotsick of Morehead State University, who burned the CD-ROM used for the data in the first edition.

People at ESRI: Jack Dangermond, who supported the idea of this text; Bill Miller, Earl Nordstrand, and Michael Phoenix, who provided encouragement and advice.

People at Trimble Navigation: Art Lange, my GPS guru, who provided considerable technical help and kept me from making several real blunders; Chuck Gilbert, Chris Ralston, and Dana Woodward, who helped by providing advice and equipment; Michele Vasquez, who provided photos for the text.

Michele Carr, of AST Research, for the use of the pen computer that facilitated GPS data collection. Carla Koford and Ethan Bond of the GIS lab at the University of Kentucky (UK), who tested and corrected procedures and text, and Jena King, who read and improved parts of the text.

Justin Stodola, who wrote the "c" program and the procedures for digitizing coverages directly into unique UTM tiles.

People who collected GPS data in faraway places: Will Holmes for the Mexican data and Chad Staddon, who took a GPS receiver to Bulgaria.

Bob Crovo, of the UK computing center, who was always cheerful about answering dumb questions. Ron Householder of MapSync and Tim Poindexter of CDP Engineers, who use GPS as professionals and know a lot that isn't in the manuals.

David Lucas, GIS coordinator for Lexington-Fayette County, who guided me through some sticky problems with UNIX and license managers and provided data on Lexington Roads.

Ken Bates—"Mr. GIS" for the state government of Kentucky—and Kent Annis of the Bluegrass Area Development District, for their help and insights.

The students in several classes of GEO 409, 506, and S09—GIS and computer-assisted cartography courses at UK—who read the book and tested the exercises.

Ruth Rowles, who used an early version of the text in her GIS class at UK.

Calvin Liu, who operates the GPS community base station at UK and provided many of the base station files used herein. And the folks at the base station in Whiterock, British Columbia, for helping a stranger with an urgent request.

Scott Samson, who provided good advice at important times.

Jon Goss and Matt McGranaghan, who facilitated my stay and lecture at the University of Hawaii—one of the nicer places to collect GPS data, or data of any sort for that matter.

Tom Poiker, of Simon Fraser University, for his help and counsel on various GIS topics, and to him and Jutta for their hospitality in their home in British Columbia—also a great place for data collection.

Max Huff, of OMNISTAR, Inc., who demystified the complex "differential corrections anywhere" system and provided hardware for same.

Several companies for hardware, software, and support: ESRI for sponsoring the project and providing software; Trimble Navigation, for providing the GPS hardware; AST Research, for use of a pen computer for collecting data while traveling; and AccuPoint and OMNISTAR for access to their differential correction services.

John Bossler, of Ohio State University, who does really high-tech GPS, and who was patient in letting me finish this text, delaying a project I was doing for him.

Hans Vinje, first officer of the Nordic Empress, who took the time to explain the way ship navigators combine GPS with other navigational systems, and to Captain Ulf Svensson, who invited me to spend time on the bridge.

The folks at Ann Arbor Press, who did indeed "press" to get the book into final form: Skip DeWall and Sharon Ray. I should add that responsibility for the appearance of the text and any errors you may find rest entirely with the author.

My colleagues in the Department of Geography at UK—and particularly to Richard Ulack, its chair—who indulged my absences and absentmindedness during the last weeks of this project.

And finally my dear friend Barb Emler, who repairs children's teeth from nine to five every day and justly believes that people ought to enjoy life without work outside of those times. She's right. And I will. For a while, anyway.

SECOND EDITION

In putting together the second edition, which for some reason was no less work than the first, I am indebted to the following people:

Dr. Thomas Meyer, assistant professor in the Department of Natural Resources Management and Engineering at the University of Connecticut, who used the text in his classes and commented on several chapters of the text, which kept me from making major mistakes in the area of geodesy.

Dr. Kenneth L. Russell, professor at Houston Community College, who inflicted draft versions of the text on his students and carefully detailed the ways in which the text worked and didn't work. His real-world experiences with GPS and GIS were immensely helpful.

Dr. Qinhua Zhang, who carefully read and tested the entire text. He painstakingly went through every word and every exercise—saving me (and you, the reader) from enduring many minor errors and a couple of major mistakes.

James Long, who kept my computing equipment humming, or at least running, despite hardware and operating system frustrations.

Teresa Di Gioia, for carefully testing two chapters.

Janice Kelly and Cooper Gillis for their hospitality in Newport, Rhode Island, which served as a base for data collection in New England. Pat Seybold, for her encouragement and assistance with development of an exercise illustrating differential correction. More people who collected GPS data in faraway places: Dr. Paul Karan for the data from Japan, and Jennifer Webster for data from Ecuador.

Terrapro in British Columbia, Canada, for differential correction files.

And some more of the helpful people at Trimble Navigation, Limited: Art Lange, as usual.

Greg May, formerly of the Geo3 team, presently involved in Trimble's precision agriculture projects, who analyzed data and answered numerous questions (after 40 e-mails and a dozen phone calls, I quit counting) on all facets of GPS.

Pat McLarin, Geo3 product manager, who helped me understand how the Geo3 is different.

Fay Davis, Pathfinder Office Product Manager, who persevered through my constant barrage of "why doesn't the software do such and such" questions.

Alan Townsend, for writing the foreword to the second edition.

Allison Walls, Andrew Harrington, Brian Gibert, Barbara Brown, Chris Ralston, Erik Sogge, Bob Morris, Paul Drummond, and Neil Briggs for important help with the myriad of little issues that crop up in writing a technical text.

THIRD EDITION

I had considerable assistance with third edition. I want to thank particularly:

Ron Bisio, of Trimble Navigation, for counseling on GPS in general and making Trimble receivers available for project development.

Robert Wick, of Magellan, for making available the MobileMapper 6 receiver and providing access to Magellan technical support.

Dan Colbert of Trimble who assisted with TerraSync, ArcMap, and ActiveSync connection problems.

Juan Luera and Todd Stuber with ESRI Technical Support, who helped me at critical times with ArcPad.

Jonathan Draffan, technical support engineer with Magellan's specializing in GIS, who unraveled some of the mysteries of ArcPad for me.

I also appreciate:

Other folks at Magellan: Joe Sass, Raphel Finelli, and Sue MacLean

Nicole Brecht and Aubrey Allmond at Trimble Navigation

Alan Cameron, Editor of GPS World, for sage words on GPS and privacy

Mark Wiljanen, of the Council of Postsecondary Education of Kentucky, and Ken Bates, again, for help in procuring the latest ESRI releases.

Damian Spangrud from ESRI for allowing me into ESRI Beta programs.

Ron Householder, GPS professional, surveyor, and Trimble Navigation representative in Lexington, Kentucky. His capable and innovative company is MapSync, Inc.

Paul Baxter, Baxter and Associates, Energy and Environmental Consulting— friend for many years who prompted me to produce this third edition.

Jeff Levy, GIS/GPS technician at the Gyula Pauer Center for Cartography & GIS at the University of Kentucky

Jill Jurgensen, Arlene Kopeloff, Amy Rodriguez, George Kenney, Irma Shagla, and Thomas Singer from Taylor & Francis, who each played a role in getting this third edition into print, and who had to abide the author's tardiness on several occasions.

And Jerri Weitzel, who is probably Kentucky's best 4th grade writing teacher, for putting up with my absences and excuses while producing this edition of GPS for GIS.

About the Author

Michael Kennedy's involvement with Geographic Information Systems (GIS) began in the early 1970s with his participation on a task force formed by the Department of the Interior to provide technical recommendations for pending federal land use legislation.

In the mid-1970s he and coauthors wrote two short books on GIS, both published by the Urban Studies Center at the University of Louisville, where he was enjoying sabbatical leave from the University of Kentucky. *Spatial Information Systems: An Introduction*, with Charles R. Meyers, is a description of the components of a GIS and was a guide to building one at the time when there was no off-the-shelf software. *Avoiding System Failure: Approaches to Integrity and Utility*, with Charles Guinn, described potential pitfalls in the development of a GIS. With Mr. Meyers and R. Neil Sampson he also wrote the chapter "Information Systems for Land Use Planning" for *Planning the Uses and Management of Land*, a monograph published in 1979 by the American Society of Agronomy.

Professor Kennedy is also a computer textbook author, having cowritten, with Martin B. Solomon, *Ten Statement Fortran Plus Fortran IV, Structured PLIZERO Plus PLIONE,* and *Program Development with TIPS and Standard Pascal*, all published by Prentice-Hall.

Over the years Professor Kennedy has had a wide range of experiences relating computers and environmental matters. Primarily to be able to talk to planners about the newly emerging field of GIS, he became certified as a planner by the American Institute of Certified Planners (AICP). He was director of the Computer-Aided Design Laboratory at the University of Kentucky for several years. He has been invited to teach GIS and/or programming at Simon Fraser University and several state or provincial universities: North Carolina, Florida, and British Columbia.

Outside of his interest in the Global Positioning System (GPS) the author's primary concern is in the development of computer data structures for the storage of geographic information. In work sponsored by the Ohio State University Center for Mapping and the Environmental Systems Research Institute, he is currently developing what he calls the dot-probability paradigm for the storage of spatial data. Fundamentally, the author is a programmer who has sought out the application of computers to environmental issues. He is currently an associate professor in the Geography Department at the University of Kentucky, where he teaches GIS and GPS.

Introduction

Two of the most exciting and effective technical developments to emerge fully in the last decade are the development and deployment of Global Positioning Systems (the GPS of the United States is called NAVSTAR) and the phenomenon of the Geographic Information System (GIS).

GIS is an extremely broad and complex field, concerned with the use of computers to input, store, retrieve, analyze, and display geographic information. Basically GIS programs make a computer think it's a map—a map with wonderful powers to process spatial information and to tell its users about any part of the world, at almost any level of detail.

Although GPS is also an extremely complex system, using it for navigation is simple by comparison. It allows you to know where you are by consulting a radio receiver. The accuracies range from as good as a few millimeters to somewhere around 15 meters, depending on equipment and procedures applied to the process of data collection.

More advanced GPS receivers can also record location data for transfer to computer memory, so GPS can tell you not only where you are but also where you were. Thus, GPS can serve as means of data input for GISs. This subject is not quite as simple as using GPS for navigation. Traditionally (if one can use that word for such a new and fast-moving technology), GISs got their data from maps and aerial photos. These were either scanned by some automated means or, more usually, digitized manually using a handheld "puck" to trace map features—the map being placed on an electronic drafting board called a "digitizer." With GPS, the earth's surface becomes the digitizer board; the GPS receiver antenna becomes the puck. This approach inverts the entire traditional process of GIS data collection: spatial data come directly from the environment and the map becomes a document of output rather than input.

A cautionary note: The aim of this text is to teach you to use GPS as a source of input to GIS. The book is somewhat unusual in that it has multiple characteristics:

It is an informational discussion, a manual, and a workbook. What I try to do is present material in a way and in an order so you can gain both obvious and subtle knowledge from it if you are paying close attention. For each major subject there is an Overview followed by Step-by-Step procedures. After each step you should think about what the step implies and what you could learn from it. As with many tutorials, it may be possible, in the early parts of the text, to go through the steps sort of blindly, getting the proper results but not really understanding the lessons they teach. I advise against that.

1 Basic Concepts

In which you are introduced to facts and concepts relating to the NAVSTAR Global Positioning System and have your first experience taking data with a GPS receiver.

OVERVIEW

A sports club in Seattle decided to mount a hunting expedition. They employed a guide who came well recommended, and whose own views of his abilities were greater still. Unfortunately, after two days, the group was completely, totally lost. "You told me you were the best guide in the State of Washington," fumed the person responsible for hiring the guide. "I am, I am," claimed the man defensively. "But just now I think we're in Canada."

A dozen years ago, when the first edition of this book appeared, stories like the one above were plausible. Now, given the right equipment* and a map of the area, you could be led blindfolded to any spot in the great out-of-doors and determine exactly where you were.† This happy capability is due to some ingenious electronics and a dozen billion dollars‡ spent by the U.S. government. I refer to NAVSTAR (NAVigation System with Time And Ranging; informally the "Navigation Star")—a constellation of from 24 to 32 satellites orbiting the Earth, broadcasting data that allows users on or near the Earth to determine their spatial positions. The more general term in the United States for such an entity is "Global Positioning System" or "GPS." The Russians have such a navigation system as well, which they call GLONASS (GLObal Navigation Satellite System). (One might reflect that, for some purposes, the cold war lasted just long enough.) A more general, recent acronym for such systems is GNSS, standing for Global Navigation Satellite Systems. In the Western world, GPS usually implies NAVSTAR, so I will use the two designations interchangeably in this text.

---

* Sales of GPS receivers have reached around nine billion annually. At, say, an average of $600 per receiver, that's a lot of units.

† On the other hand, no one is quite as potentially lost as the person who depends solely on a GPS unit that runs out of electricity. Maps don't need batteries!

‡ But, as you will see soon, we get a lot for this $12 billion. As this book goes to press the U.S. government is in the process of spending many times that much for an antiballistic missile system (ABM) which is highly unlikely to be able to stop a missile attack and guaranteed not to stop an attack that comes by truck, small boat, or other conveyance. In other words, although GPS was expensive, it does have the advantage of working—something that ABM will not do.

WHERE ARE YOU?

Geography, and Geographic Information Systems (GIS) particularly, depends on the concept of location. Working with "location" seems to imply that we must organize and index space. How do we do that?

Formally, we usually delineate geographical space in two dimensions on the Earth's surface with the latitude-longitude graticule, or with some other system based on that graticule.

But informally, and in the vast majority of instances, we organize space in terms of the features in that space. We find a given feature or area based on our knowledge of other features—whether we are driving to Vancouver or walking to the refrigerator. Even planes and ships using radio navigational devices determine their positions relative to the locations of fixed antennae (though some of the radio signals may be converted to graticule coordinates).

Unlike keeping track of time, which was initially computed relative to a single, space-based object (the Sun), humans kept track of space—found their way on the ground—by observing what was around them.

Another, somewhat parallel way of looking at this issue is in terms of absolute versus relative coordinates. If I tell you that Lexington, Kentucky, is at 38 degrees (38°) north latitude, 84.5° west longitude, I am providing you with absolute coordinates. If I say, rather, that Lexington is 75 miles south of Cincinnati, Ohio, and 70 miles east of Louisville, Kentucky, I have given you relative coordinates.

Relative coordinates usually appeal more to our intuitive comprehension of "location" than do absolute coordinates; this does not mean, however, that relative coordinates cannot be quite precise.

To pass spatial information around, humans developed maps to depict mountains and roads, cities and plains, radio stations and sinkholes. Maps aid both the formal and informal approaches that humans use to find objects and paths. Some maps have formal coordinates, but maps without graticule markings are common. All maps appeal to our intuitive sense of spatial relationships. The cartographer usually relies on our ability to use the "cognitive coordinates" in our memory, and our abilities to analyze, to extrapolate, and to "pattern match" the features on the map. It is good that this method works, because, unlike some amazing bird and butterfly species, humans have no demonstrated sense of an absolute coordinate system. But with maps, and another technological innovation, the magnetic compass, we have made considerable progress in locating ourselves.

I do not want to imply that absolute coordinates have not played a significant part in our position-finding activities. They have, particularly in navigation. At sea, or flying over unlit bodies of land at night, captains and pilots used methods that provided absolute coordinates. One's position, within a few miles, can be found by "shooting the stars" for a short time with devices such as sextants or octants. So the GPS concept—finding an earthly position from bodies in space—is not an entirely new idea. But the ability to do so during the day, almost regardless of weather, with high accuracy and almost instantaneously, makes a major qualitative difference. As a parallel, consider that a human can move by foot or by jet plane. They are both methods of locomotion, but there the similarity ends.

GPS, then, gives people an easy method for both assigning and using absolute coordinates. Now, humans can know their **positions** (i.e., the coordinates that specify where they are); combined with map and/or GIS data they can know their **locations** (i.e., where they are with respect to objects around them). I hope that, by the time you've completed this text and experimented with a GPS receiver, you will agree that NAVSTAR constitutes an astounding leap forward.

WHAT TIME IS IT?

Although this is a text on how to use GPS in GIS (ArcGIS in particular)—and hence is primarily concerned with positional issues, it would not be complete without mentioning what may, for the average person, be the most important facet of GPS: providing Earth with a universal, exceedingly accurate time source. Allowing any person or piece of equipment to know the exact time has tremendous implications for things we depend on every day (like getting information across the Internet, like synchronizing the electric power grid and the telephone network). If we were naming the system today, knowing what we know now, we would probably call it GPTS: the Global Positioning and Timing System. An added benefit is that knowing time so precisely allows human knowledge to be enhanced by research projects that depend on knowing the exact time in different parts of the world. For example, it is now possible to track seismic waves created by earthquakes, from one side of the Earth, through its center, to the other side, because the **exact** time* may be known worldwide.†

GPS AND GIS

The subject of this book is the use of GPS as a method of collecting locational data for Geographic Information Systems (GIS). We will concentrate on using ArcGIS from ESRI in Redlands, California.‡ The appropriateness of this seems obvious, but let's explore some of the main reasons for making GPS a primary source of data for GIS:

- Availability: In 1995, the U.S. Department of Defense (DoD) declared NAVSTAR to have "final operational capability." Deciphered, this means that the DoD has committed itself to maintaining NAVSTAR's capability for civilians at a level specified by law, for the foreseeable future, at least in times of peace. Therefore, those with GPS receivers may locate their positions anywhere on the Earth.
- Accuracy: GPS allows the user to know position information with remarkable accuracy. A receiver operating by itself can let you locate yourself within 5 to 10 meters of the true position. (And later you will learn how to get accuracies of 2 to 5 meters.) At least two factors promote such accuracy:

---

* Well, right, there is no such thing as "exact" time. Time is continuous stuff like position, speed, water and real numbers, not discrete stuff like people, eggs, cars, and integers, so when I say exact here I mean within a variation of a few billionths of a second—a few nanoseconds.
† The baseball catcher Yogi Berra was once asked "Hey Yogi, what time is it?" to which Berra is said to have replied, "You mean right now?" Yes, Yogi, **RIGHT NOW!**
‡ http://www.esri.com.

- First, with GPS, we work with primary data sources. Consider one alternative to using GPS to generate spatial data: the digitizer. A digitizer is essentially an electronic drawing table, wherewith an operator traces lines or enters points by "pointing"—with "crosshairs" embedded in a clear plastic "puck"—at features on a map.

- One could consider that the ground-based portion of a GPS system and a digitizer are analogous: the Earth's surface is the digitizing tablet, and the GPS receiver antenna plays the part of the crosshairs, tracing along, for example, a road. But data generation with GPS takes place by recording the position on the most fundamental entity available: the Earth itself, rather than a map or photograph of a part of the Earth that was derived through a process involving perhaps several transformations.

- Second, GPS itself has high inherent accuracy. The precision of a digitizer may be 0.1 millimeters (mm). On a map of scale 1:24,000, this translates into 2.4 meters (m) on the ground. A distance of 2.4 m is comparable to the accuracy one might expect of the properly corrected data from a medium-quality GPS receiver. It would be hard to get this out of the digitizing process. A secondary road on our map might be represented by a line five times as wide as the precision of the digitizer (0.5 mm wide), giving a distance on the ground of 12 m, or about 40 feet.

- On larger-scale maps, of course, the precision one might obtain from a digitizer can exceed that obtained from the sort of GPS receiver commonly used to put data into a GIS. On a "200-scale map" (where 1 inch is equivalent to 200 feet on the ground) 0.1 mm would imply a distance of approximately a quarter of a meter, or less than a foot. Although this distance is well within the range of GPS capability, the equipment to obtain such accuracy is expensive and is usually used for surveying, rather than for general GIS spatial analysis and mapmaking activities. In summary, if you are willing to pay for it, at the extremes of accuracy, GPS wins over all other methods. Surveyors know that GPS can provide horizontal, real-world accuracies of less than 1 centimeter.

- Ease of use: Anyone who can read coordinates and find the corresponding position on a map can use a GPS receiver. A single position so derived is usually accurate to within 7 meters or so. Those who want to collect data accurate enough for a GIS must involve themselves in more complex procedures, but the task is no more difficult than many GIS operations.

- GPS data are inherently three-dimensional: In addition to providing latitude-longitude (or other "horizontal" information), a GPS receiver may also provide altitude information. In fact, unless it does provide altitude information itself, it must be told its altitude in order to know where it is in a horizontal plane. The accuracy of the third dimension of GPS data is not as great as the horizontal accuracies. As a rule of thumb, variances in the horizontal accuracy should be multiplied by 1.5 (and perhaps as much as 3.0) to get an estimate of the vertical accuracy.

Anatomy of the Term "Global Positioning System"

Global: Anywhere on Earth. Well, almost anywhere, but not (or not as well)

- inside buildings,
- underground,
- in very severe precipitation,
- under heavy tree canopy,
- around strong radio transmissions,
- in "urban canyons" amongst tall buildings,
- near to powerful radio transmitter antennas,

or anywhere else not having a direct view of a substantial portion of the sky. The radio waves that GPS satellites transmit have very short lengths—about 20 cm. A wave of this length is good for measuring because it follows a very straight path, unlike its longer cousins, such as AM and FM band radio waves, which may bend considerably. Unfortunately, short waves also do not penetrate matter very well, so the transmitter and the receiver must not have much solid matter between them, or the waves are blocked, as light waves are easily blocked.

Positioning: Answering brand-new and age-old human questions. Where are you? How fast are you moving and in what direction? What direction should you go to get to some other specific location, and how long would it take at your speed to get there? **And, most importantly for GIS, where have you been?**

System: A collection of components with connections (links) among them. Components and links have characteristics. GPS might be divided up in the following way:*

The Earth

The first major component of GPS is the Earth itself: its mass and its surface, and the space immediately above. The mass of the Earth holds the satellites in orbit. From the point of view of physics, each satellite is trying to fly by the Earth at 4 kilometers per second. The Earth's gravity pulls on the satellite vertically so it falls. The trajectory of its fall is a track that is parallel to the curve of the Earth's surface.

The surface of the Earth is studded with little "**monuments**"—carefully positioned metal or stone markers—whose coordinates are known quite accurately. These lie in the "numerical **graticule**" which we all agree forms the basis for geographic position. Measurements in the units of the graticule, and based on the positions of the monuments, allow us to determine the position of any object we choose on the surface of the Earth.

---

* Officially, the GPS system is divided up into a Space Segment, a Control Segment, and a User Segment. We will look at it a little differently. One of the many places to see official terminology—admittedly somewhat dated—is http://tycho.usno.navy.mil/gpsinfo.html#seg.

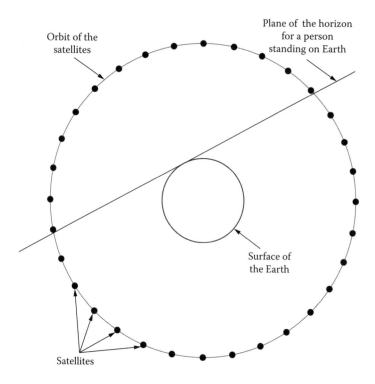

FIGURE 1.1   Satellites visible from a position on Earth.

Earth-Circling Satellites

The United States GPS design calls for a total of at least 24 and up to 32 solar-powered satellite radio transmitters, forming a constellation such that several are "visible" from any point on Earth at any given time. Figure 1.1 indicates the maximum number of satellites that would be visible from horizon to horizon. You will rarely be able to detect that many for a variety of reasons. The first one was launched on February 22, 1978. In mid-1994 24 were broadcasting. The minimum "constellation" of 24 includes three "spares." As many as 31 have been up and working at one time. The satellites are also referred to as Space Vehicles or SVs. The newest SVs are designated as type Block II-R-M.

The GPS satellites (called SVs) are at a "middle altitude" of about 11,000 nautical miles (nm), or roughly 20,400 kilometers (km) or 12,700 statute miles (sm) above the Earth's surface. This puts them considerably above the standard orbital height of the space shuttle, most other satellites, and the enormous amount of space junk that has accumulated. They are also well above Earth's air, where they are safe from the effects of atmospheric drag. When GPS satellites "die" they are sent to orbits about 600 miles farther out.

GPS satellites are considerably below the geostationary satellites, which are usually used for communications and sending TV, telephone, and other signals back to Earth-based fixed antennas. These satellites are 35,763 km (or 19,299 nm or 22,223

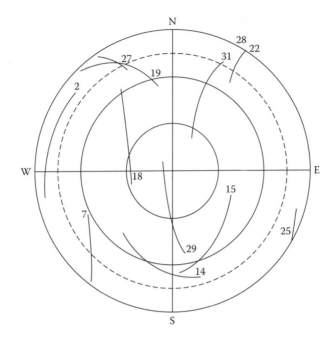

FIGURE 1.2 GPS Satellite tracks looking from space toward the Equator.

sm) above the Earth, where they hang over the equator relaying signals from and to ground-based stations. (The geo-stationary satellites have to be above the equator to stay in synchronization with Earth's rotation. And they have to be synchronized with Earth's rotation; otherwise you couldn't point satellite dishes at them.)

The NAVSTAR satellites are neither polar nor equatorial, but slice the Earth's latitudes at about 55°, executing a single revolution every 12 hours. Further, although each satellite is in a 12-hour orbit, an observer on Earth will see it rise and set about 4 minutes earlier each day.* There are four or five satellites in slots in each of six distinct orbital planes (labeled A, B, C, D, E, and F) set 60 degrees apart. The orbits are almost exactly circular.

The combination of the Earth's rotational speed and the satellites' orbits produces a wide variety of tracks across the Earth's surface. Figure 1.2 is a view of the tracks which occurred during the first two hours after noon on St. Patrick's Day, 1996. You are looking down on the Earth, directly at the equator and at a (north-south) meridian that passes through Lexington, Kentucky. As you can see by Figure 1.2, the tracks near the equator tend to be almost north-south. The number of each satellite is shown near its track; the number marks the point where the satellite is at the **end** of the two-hour period.

GPS satellites move at a speed of 3.87 km/sec (8,653 miles per hour). The Block IIR satellites weigh about 1077 kilograms (somewhat more than a ton) and have a length of about 11.6 meters (about 38 feet) with the solar panels extended. Those

---

* Why? Answers to several questions that may occur to you will be supplied in Chapters 2 and 3. We avoid the digression here.

FIGURE 1.3   A NAVSTAR GPS satellite.

panels generate about 1100 watts of power. The radio on board broadcasts with about 40 watts of power. (Compare that with some FM stations with 50,000 watts.) The radio frequency used for the civilian GPS signal is called "GPS L1" and is at 1,575.42 megahertz (MHz). Space buffs might want to know that the SVs usually get into orbit on top of Boeing Delta II rockets fired from the Kennedy Spaceflight Center at Cape Canaveral in Florida. Each satellite has on board four atomic clocks (either cesium or rubidium) that keep time to within a billionth of a second or so, allowing users on the ground to determine the current time to within about 40 billionths of a second (40 nanoseconds). Each satellite is worth about $65 million and has a design life of 10 years. Figure 1.3 shows an image of a NAVSTAR satellite.

Ground-Based Stations

Although the GPS satellites are free from drag by the atmosphere, their tracks are influenced by the gravitational effects of the Moon and Sun, and by the solar wind. Further, they are crammed with electronics. Thus, both their tracks and their innards require monitoring. This is accomplished by five ground-based stations near the equator, located on Ascension Island (in the South Atlantic), at Diego Garcia (in the Indian Ocean), on Kwajalein Atoll and in Hawaii (both in the Pacific), and at Cape Canaveral, Florida, plus the master control station (MCS) at Schriever (formerly Falcon) Air Force Base near Colorado Springs, Colorado. Each satellite passes over at least one monitoring station twice a day. Information developed by the monitoring station is transmitted back to the satellite, which in turn rebroadcasts it to GPS receivers. Subjects of a satellite's broadcast are the health of the satellite's electronics, how the track of the satellite varies from what is expected, the current **almanac*** for all the satellites, and other, more esoteric subjects which need not concern us at this point. Other ground-based stations exist, primarily for uploading information to the satellites.

Receivers

This is the part of the system with which you will become most familiar. In its most basic form, the satellite receiver consists of

- An antenna (whose position the receiver reports)
- Electronics to receive the satellite signals
- A microcomputer to process the data and to record position values
- Controls to provide user input to the receiver
- A screen to display information

The elaborate units have computer memory to store position data points and the velocity of the antenna. This information may be uploaded into a personal computer or workstation and then installed in a GIS software database. Another elaboration on the basic GPS unit is the ability to receive data from and transmit data to other GPS receivers—a technique called "real-time differential GPS" that may be used to considerably increase the accuracy of position finding.

Receiver Manufacturers

In addition to being an engineering marvel and of great benefit to many concerned with spatial issues as complex as national defense or as mundane as refinding a great fishing spot, GPS is also big business. Dozens of GPS receiver builders exist—from those who manufacture just the GPS "engine," to those who provide a complete unit for the end user. In this text we explain the concepts in general but use equipment manufactured by Trimble Navigation, Ltd., and Magellan, Inc., because their units work well, are quite accurate, have programs of educational discounts, and are likely to be part of educational GPS labs throughout the world.

The United States Departments of Defense and Transportation

The U.S. DoD is charged by law with developing and maintaining NAVSTAR. It was, at first, secret. Five years elapsed from the first satellite launch in 1978 until news of GPS came out in 1983. The story, perhaps apocryphal, is that President Reagan, at the time a Korean airliner strayed into Soviet airspace and was shot down, lamented something like, "This wouldn't have happened if the damn GPS had been up." A reporter who overheard wanted to know what GPS was. In the almost three decades since—despite the fact that parts of the system remain highly classified—mere citizens have been cashing in on what Trimble Navigation, Ltd. calls "The Next Utility." Probably the military planners of those days did not envision that an activity called geocaching would have more than 5 million pages on the Internet, that automobile

---

* An almanac is a description of the predicted positions of heavenly bodies.

navigation systems and location-based services would be big businesses, or that GPS would become a primary method for mapmaking and spatial analysis.

There is little question that the design of GPS would have been different had it been a civilian system "from the ground up." But then, GPS might not have been developed at all. Many issues must be resolved in the coming years. A Presidential Directive issued in March of 1996 designated the U.S. Department of Transportation as the lead civilian agency to work with DoD so that nonmilitary uses can bloom. DoD is learning to play nicely with the civilian world. They and we all hope, of course, that the civil uses of GPS will vastly outpace the military need.

One important matter has been addressed: For years the military deliberately corrupted the GPS signals so that a single GPS unit, operating by itself (i.e., **autonomously**), could not assure accuracy of better than 100 meters. This policy (known as Selective Availability (SA)) was terminated on 2 May 2000. Now users of autonomous receivers may know their locations within about 5 meters.

Users

Finally, of course, the most important component of the system is **you**: the "**youser**," as my eight-year-old spelled it. A large and quickly growing population, GIS users come with a wide variety of needs, applications, and ideas. From tracking ice floes near Alaska to digitizing highways in Ohio. From rescuing sailors to pinpointing toxic dump sites. From urban planning to forest management. From improving crop yields to laying pipelines. Welcome to the exciting world of GPS!

How We Know Where Something Is

First, a disclaimer: this text does not pretend to cover issues such as geodetic datums, projections, coordinate systems, and other terms from the fields of geodesy and surveying. In fact, I am going to make it a point not even to define most of these terms, because simply knowing the definitions will not serve you unless you do a good bit of study. Many textbooks and Web pages are available for your perusal. These fields, concepts, and principles may or may not be important in the collection of GPS information for your GIS use—depending on the sort of project you undertake. What is important, vital in fact, is that when GPS data is to be combined with existing GIS or map information, the datum designation, the projection designation, the coordinate system designation, and the measurement units that are used must be identical.

Before we undertake to use a GPS receiver to determine a **position**, it is important to understand what is meant by that term. It seems like a straightforward idea, but it has confused a lot of people, particularly when a given position is described as a set of numbers.*

Take a tent stake, or a knitting needle, or a table knife and stick it vertically into a patch of ground. Now reflect upon the fact that the object is not moving with respect

---

* Descriptions of points aren't always numbers. A possibly apocryphal "meets & bounds" description a couple of centuries ago of a point in Kentucky was "Two tomahawk throws from the double oak in a northerly direction."

to the planet,* as it would be if you stuck it into the soil in a flowerpot on the deck of a ship. The location of the object may be identified through time by three unchanging numbers of latitude, longitude, and altitude. In other words, it is where it is. But in the last century, teams of mathematicians and scientists (skilled in geodesy) have developed other sets of numbers to describe exactly the same spot where your object now resides. **The spot's position doesn't change, but the description of it does.**

Ignoring the matter of altitude for the moment, suppose you put your object into the ground at latitude 38.0000000° (North) and longitude 84.5000000° (West), according to the latest belief about where the center of the Earth is and where its poles are. This most recent, widely accepted view is the datum described as the World Geodetic System of 1984 (WGS84), based on the GRS80 ellipsoid.† In the coterminous states of the United States, this datum is said to be virtually identical (within millimeters) with the North American Datum of **1983** (NAD83). As I indicated, if you want to know more there are plenty of sources.

Prior to this latest determination of the latitude-longitude graticule, many people and organizations in the United States used the North American Datum of **1927** (NAD27) as the best estimate of where the latitude-longitude graticule was positioned, based on parameters determined by Clarke in 1866. According to this datum, your object would be at latitude 37.99992208° and longitude 84.50006169°. This looks like an insignificant difference but translates into about 10 meters on the Earth's surface. To consider another way of thinking about what this means: If you put a second object in the ground at 38° latitude and 84.5° according to NAD27, it would be 10 meters away from the first one. Does that sound like a lot? People have been shot in property disputes over much smaller distances. So, when a GPS receiver gives you a latitude and longitude you must know the datum that is the basis for the numbers. There are hundreds of datums—many countries have their own. NAVSTAR GPS is based fundamentally on WGS84 but lots of GPS units, including the ones you will be using, can calculate and display coordinates in many other datums.

There are many reasons that people do not find it convenient to use latitude and longitude to describe a given point, or set of points, on the Earth's surface. One is that doing calculations using latitude and longitude—say determining the distance between two points—is a pretty complex matter involving such things as products of sines and cosines. For a similar distance calculation, if the points are on the Cartesian plane, the worst arithmetic hurdle is a square root.

Another reason not to use latitude and longitude measures for many applications is that it doesn't work well for several aspects of map making. Suppose you plot many points on the Earth's surface—say the coastline of a small island some distance from the equator—on a piece of ordinary graph paper, using the longitude numbers for "x" coordinates and latitude numbers for "y" coordinates. The shape of the island might look pretty strange compared with how it would appear if you got

---

* Well, it's not moving much. If you are in Hawaii it is moving northwest at about 4 inches per year. If you are unfortunate enough to do this exercise when there is an earthquake happening it might be moving quite a bit, and it might not return to its original position.

† GRS80 is a global geocentric system based on the ellipsoid adopted by the International Union of Geodesy and Geophysics (IUGG) in 1979. GRS80 is the acronym for the Geodetic Reference System 1980.

up in an airplane and looked down on it. And if you measured distances or angles or areas on the plot you would not get useful numbers. This is due to a characteristic of the spherical coordinate system: the length of an arc of a degree of longitude does not equal the length of an arc of a degree of latitude. It comes close to equal near the equator but it is nowhere near close as you go further north or south from the equator. At the equator a degree of longitude is about 69.17 miles. Very near the North Pole a degree of longitude might be 69.17 inches. (A degree of latitude, on the other hand, varies only between about 68.71 miles near the equator and 69.40 miles near the poles.)

A good solution to these problems (calculation and plotting) for relatively small areas is the concept of a "projection" with which you are no doubt familiar. The term comes from imagining a process in which you place a light source within a transparent globe that has features inscribed on it, and let the light fall on a flat piece of paper (or one that is curved in only one direction and may be unrolled to become flat).* The shadows of the features (say lines, or areas) will appear on the paper. You can then apply a Cartesian coordinate system to the paper, which gives you the advantages of easy calculation and more realistic plotting. However, distortions are inherent in any projection process; most of the points on the map will not correspond exactly to their counterparts on the ground. The degree of distortion is greater on maps that display more area.

In summary, latitude and longitude numbers of a given datum provide an exact method of referencing any given single point, but they are difficult to calculate with, and multiple points suffer from distortion problems when plotted. Projected coordinate numbers are easy to calculate with but, in general, misplace points with respect to other points—thus producing errors in the distances, sizes, shapes, and/ or directions.

Let's look at some other common representations of the position of our object.

A coordinate system called Universal Transverse Mercator (UTM) was developed based on a series of 60 transverse Mercator projections. These projections are further subdivided into areas, called "**zones**," covering 6° of longitude and, usually, 8° of latitude. A coordinate system is imposed on the resulting projection such that the numbers are always positive "to the right and up."

The representation of our object (at 38°N and 84.5°) in the UTM coordinate system, when that system is based on WGS84, is a "**northing**" of 4,208,764.4636 meters and an "**easting**" of 719,510.3358 meters. The northing is the distance along the surface of a "bumpless" Earth, in meters, from the equator. The easting is somewhat more complicated to explain since it depends on the zone and a coordinate system that allows the number to always be positive. Consult a textbook on geodesy, cartography, or look at some of the ten thousand plus Web pages that come up when you type "UTM coordinate system transverse Mercator" into an Internet search engine (e.g., http://www.google.com).

---

* This is a sort of a cartoon of what a map projection is. A map projection is actually a mathematical transformation that "maps" points on the globe to points on a plane; depending on the projection, the process is quite complex; a single light source at the center of the globe does not suffice to explain it.

However, there is also a version of the UTM coordinate system that is based on NAD27. In this case, our object would have different coordinates: Northing 4,208,550.0688 and Easting 719,510.6393, which makes for a difference of about 214 meters. If you compare these coordinates with the previous UTM coordinates you see that virtually all of the difference is in the north-south direction. While this is true for this particular position, it is not true in general. Obviously if you tried to combine WGS84 UTM data with NAD27 UTM data the locations they depict would not match.

All U.S. states have State Plane Coordinate Systems (SPCSs) developed by the U.S. Coast and Geodetic Survey, originally in the 1930s. These systems—sometimes two or more are required for a given state—are based on different projections (mostly, transverse Mercator or another called Lambert Conformal Conic) depending on whether the state is mostly north-south (like California) or mostly east-west (Tennessee). The units are either feet or meters. Zone boundaries frequently follow county boundaries. As you can infer, each state operates pretty independently and the coordinate system(s) used in one state are not applicable in its neighbors. In Kentucky's North Zone 38.0000000° (North) and longitude 84.5000000° (West) would translate into a northing of 1,568,376.1900 feet and an easting of 182,178.3166 feet when based on WGS84. However, when the basis is NAD27, the coordinates are 1,927,939.8692 and 182,145.9821, which makes a difference of some 68 miles!

Why the large differences in coordinate systems based on NAD27 versus those based on WGS84? Because those responsible for the accuracy of other coordinate systems took good advantage of the development of WGS84—a worldwide, Earth-centered, latitude-longitude system—to make corrections or improvements to those other systems.

State plane coordinate systems have a scale error maximum of about 1 unit in 10,000. Suppose you calculated the Cartesian distance (using the Pythagorean theorem) between two points represented in a state plane coordinate system to be exactly 10,000 meters. Then, with a perfect tape measure pulled tightly across an idealized planet, you would be assured that the measured result would differ by no more than 1 meter from the calculated one. The possible error with the UTM coordinate system may be larger: 1 in 2,500.

STEP-BY-STEP

Disclaimer

The Step-by-Step sections of this book depend on hardware, firmware, and software created by various manufacturers. As you know, such entities may change over time, slightly or radically, as bugs are found, improvements are made, or for other reasons. All the Step-by-Step procedures were checked out in the summer of 2008 and found to work. The disclaimer is: there is no guarantee that they will work for you exactly as intended. You may have to be flexible and look for workarounds or find different menu items. The help files may help; they may not. But, anyway, you wouldn't be trying to use computers if you weren't flexible, would you?

FIRST OFF

Identify the computer you are going to use for this course. Make a folder, preferably under the computer's root folder C:\, named MyGPS_yis, where yis stands for YourInitialS. Substitute your initials. (My folder, for example, would be MyGPS_MK.) Into this folder will go your FastFactsFile (a text document holding information about GPS details, which I'll explain in detail later), and the data you collect. You can also put there the answers to the blanks in the textbook; the questions, including page and step numbers, are on the DVD that comes with the text.

DIFFERENT RECEIVERS AND DIFFERENT SOFTWARE

You may be using a Trimble Juno GPS receiver. You may be using a Magellan MobileMapper 6 (MM6) receiver. You may be using a receiver that is different from either of these. What I will do in the Step-by-Step sections of this text is to describe an operation in general terms and then follow it with specific instructions for the Juno receiver, sometimes with instructions for the MM6, and sometimes with a blank box indicating that you will have to come up with the procedure by reading the appropriate help file or manual for your receiver, or by using the material your instructor provides.

In the three boxes that follow you will find instructions for different equipment or software. For example:

Information about the Juno ST from Trimble Navigation will appear in this double line sort of enclosure. The Juno ST is manufactured by Trimble Navigation, Ltd., of Sunnyvale, California. They make many other GPS receivers, including "high-end" precise units used for surveying. The Juno ST is the smallest and lightest of the Juno Series of receivers. Trimble has introduced the Juno SB and SC. These units have twice the RAM and processing speed, a larger display, full SBAS support, and an image capture camera. While most of the discussion in the text relates to the Juno ST it will also apply to the SB and SC, but if you are using one of these units you should refer to the specific operating manual. (You should, of course, use the latest version of TerraSync that is available.)

Information about the MobileMapper 6 from Magellan, Inc. would appear in this single line sort of enclosure. The MobileMapper is manufactured by Magellan of Santa Clara, California. They make many other GPS receivers, including those for aircraft. They also make GPS software specific to the MMG and for PCs.

This dotted line enclosure would indicate that you should go to C:\MyGPS where you are keeping your FastFactsFile "book" on GPS and enter the information, indicating the Chapter and Step number, for your particular unit or software.

Above each set of boxes you will find a step number followed by general text describing the operation to perform.

PREPARATION

For the Step-by-Step part of this chapter, you should obtain a topo map (a USGS topographic quadrangle if you are in the United States) of the area in which you are planning to collect data. This map will also indicate the Universal Transverse Mercator (UTM) zone (e.g., 16), which you should write down for use in PROJECT 1C. Your UTM Zone is _____.

Also, see if you can find a nearby survey marker with known coordinates. (A Web site listing all U.S. Coast and Geodetic Survey markers may be found at http://www.ngs.noaa.gov.* Click on the hyperlink that invites you to get the data sheets related to "Find a point.") Also many municipalities, regional planning agencies, and state governments place monuments.

PROJECT 1A

Getting Acquainted with a GPS Receiver

{__1}† You begin your first Global Positioning System project by becoming acquainted with a typical GPS receiver, while still inside a building. Your investigation begins with an open desktop or laptop computer file (MyGPS), a notebook, and one of the following receivers:

> **Juno ST:** The Juno, with its built-in antenna, is a handheld device about 2.4" wide by 4.4" long and less than 0.8" thick; it weighs about 5 ounces. It is powered in the field by an internal lithium-ion battery or, in an automobile or boat, through a 12-volt "cigarette lighter" adapter. The AC current charger can accept 100- to 240-volt input and comes with a variety of adapter outlets in different parts of the world. An external antenna is available. Alternative power sources are "AA" batteries and the auxiliary power (cigarette lighter) receptacles in a car, boat, or airplane. For detailed information, look at http://www.trimble.com/junost.shtml.

> **MobileMapper 6:** The MM6 is a handheld device and has a built-in antenna. It is a rugged unit, measuring in inches 5.7 by 2.6 by 1.1, that weighs a little less than half a pound. The power supply is two "AA" size regular or rechargeable batteries. It contains a camera which can photographically document sites of GPS readings. An external antenna is available. For more information, go to: ftp://ftp.magellangps.com/MobileMapping

---

* http://www.ngs.noaa.gov takes you to the Web site of the National Geodetic Survey, which is part of the National Oceanic and Atmospheric Administration.

† The designation "__nn" indicates a step number, which implies that some action is required on your part. Perhaps you will want to check off each activity as it is performed. If the sentence following "__nn" is in **italics,** that indicates a general activity that is to occur; that activity is further explained by succeeding statements, so read it all before you begin the step.

> See the various manuals for your GPS receiver. Record the information in your FastFactsFile in your MyGPS folder.

The notebook—with paper and a sturdy writing surface—is there partly to get you used to the idea that a GPS receiver and a note taking ability must go hand-in-hand. After this first project, most of the data you take will be recorded in computer files, but some will not and must be written down. The notebook will also provide an index to the computer files you record.

A GPS unit is very complex internally, containing signal reception electronics and a microcomputer. With the Juno ST and the MobileMapper the GPS function is embedded within a handheld computer with a considerable memory, operating under the Microsoft Windows Mobile operating system (OS). So you not only have to learn to operate the GPS software (we illustrate TerraSync and ArcPad) but you have to know something about the OS. There isn't enough space in this book to give more than the most cursory look at the OS, but we will touch on it enough to get you to the GPS software.

Juno ST: Let's begin by looking at the hardware and descriptions of the various user interface components. See Figure 1.4 for a photo of the Juno ST.

Front View of the Juno ST GPS receiver from Trimble Navigation, Ltd.

Left, right, and back views of the Juno ST are in Figures 1.5, 1.6, and 1.7.

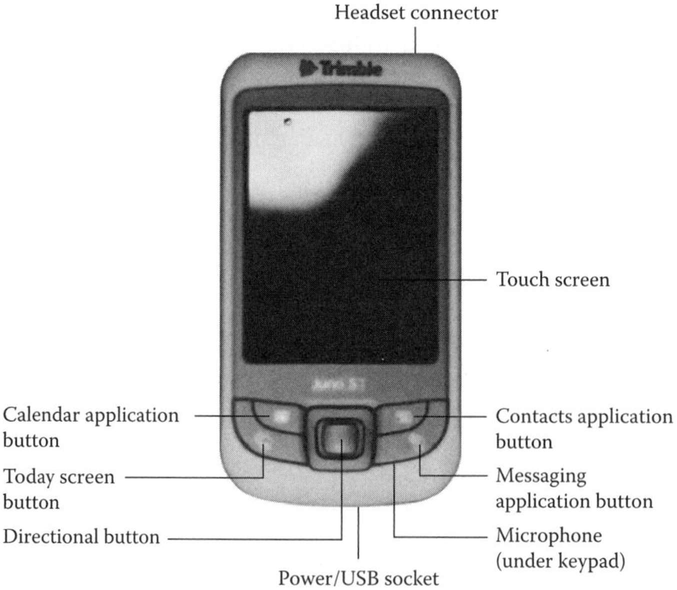

FIGURE 1.4   Front view of Trimble Juno ST receiver.

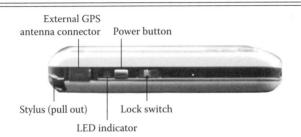

FIGURE 1.5   Left side view of Juno receiver.

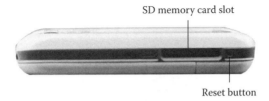

FIGURE 1.6   Right side view of Juno receiver.

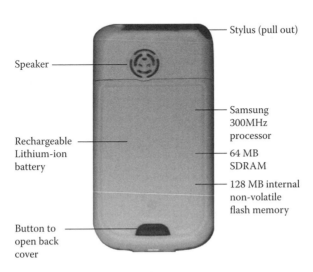

FIGURE 1.7   Back view of Juno receiver.

MobileMapper 6: Let's begin by looking at the hardware and descriptions of the various user interface components. See Figure 1.8 for a photo of the MobileMapper 6. Table 1.1 contains descriptions of MM6 elements.

FIGURE 1.8   Front view of Magellan MobileMapper 6 receiver.

TABLE 1.1

Item	Description	Item	Description
1	2.7-inch color touch screen	11	Battery compartment
2	Keypad (area not identified)	12	Battery access latch
3	Enter button	13	Backlight and flashlight button
4	Scroll button	14	Hold (freeze buttons) switch
5	Stylus and Stylus holder	15	External antenna input
6	GPS Antenna	16	Power button
7	Microphone	17	Access to SD card slot
8	Camera lens for stills and video	18	Earphone output
9	Loudspeaker	19	Flashlight light source
10	USB cable connection	20	Windows Mobile reboot button

Left, right, back, and top views of the MobileMapper 6 are in Figures 1.9, 1.10, 1.11, and 1.12.

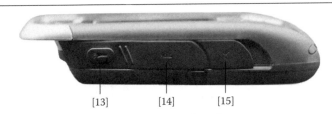

[13]   [14]   [15]

FIGURE 1.9   Left side view of MobileMapper 6 receiver.

[18]   [17]   [16]

FIGURE 1.10   Right side view of MobileMapper 6 receiver.

Rear View

[8]   [9]

[20]

[10]   [11]   [12]   [19]

FIGURE 1.11   Back view of MobileMapper 6 receiver.

FIGURE 1.12   Top view of MobileMapper 6 receiver.

Explore the user manuals for the GPS unit you are using for photos and designation of the various control buttons.

Power On and Off

{__2} *Determine how to turn your receiver on (and off). Maybe you should do a couple of cycles of this.*

Juno ST: Find the stylus on the left side of the receiver and extract it. To turn on the Juno ST press the silver button on the left upper side. If it does not come on make sure the silver slide just below it is in the "up" (unlocked) position. The screen should say Start at the top and you should see the Microsoft "flag." (If it doesn't, click any "X" icons or OK icons to terminate any running programs.) Use the stylus to touch Start. Then touch Today (which is sort of a "home" position for the Microsoft Windows Mobile operating system software). Push the silver button again to turn off the unit. Replace the stylus!

MobileMapper 6: Find the stylus on the bottom of the receiver and extract it. To turn on the MobileMapper 6 press the On/Off button on the right side for two seconds or so. After a brief pause the screen will become white, as the unit prepares to go through its startup routine. Shortly the screen should say Start at the top and you should see the Microsoft "flag." (If it doesn't, click any "X" icons, OK icons, or tap the screen to terminate any running programs.) Use the stylus to touch Start. Then touch Today (which is sort of a "home" position for the Microsoft Windows Mobile operating system software.

   To turn the unit completely off press and hold the On/Off button either very briefly or for a full five seconds. If you press it for a second or two you may simply turn off the illuminated screen. If you detect a "ghost" of information on the display the unit is still on and will continue to use up the batteries.

> Practice turning your GPS unit on and off.

I do not wish to be a pessimist, but each of these handheld receivers, operating as they do under the Microsoft Mobile operating system, sometimes go all catawampus. If your receiver freezes up or acts strangely, you may need to do a reset. You should try other solutions first, but if you have to reset here's how:

Juno ST: To reset just the GPS receiver from inside TerraSync use Setup > Options > Reset GPS receiver. If this doesn't fix things try a soft reset: Using the stylus, press and release the Reset button found in a hole on the upper right side of the receiver. If the problem persists do a hard reset: Hold in the chrome on-off button. Press and release the Reset button, then release the on-off button.

MobileMapper 6: Take the stylus and press in the Reset button found at the top of the unit. If you do a reset you may have to also reinput the date and time.

Your GPS: Consult the manual. Record the information in your FastFactsFile in your MyGPS folder.

{__3} *You can adjust the brightness of the screen (and therefore the power consumption and battery life).*

Juno ST: Press and hold the power button until the screen dims. Press and hold it again until it brightens.

MobileMapper 6: Repeatedly press the flashlight button until you get the screen intensity you want—keeping in mind that the more light you see the more often you need to replace batteries.

See the various manuals for your GPS receiver. Record the information in a file in your MyGPS folder.

Getting Power to the GPS Receiver

Most handheld devices have several options for operating and/or charging the batteries on which they operate. When they are connected to an operating computer through a USB cable they get (some) electricity that way, which can power them and (somewhat) charge the batteries if they are of the rechargeable type.

Another option for operating some receivers is to plug the power cord into a car, boat, or aircraft DC power outlet. The connector is designed to fit a standard cigarette lighter. The acceptable range of voltage input is usually broad: from 9 to 32 volts. While this is a good way to power the receiver when such power sources are available, you must be careful not to start or stop the engine of the vehicle while the unit is plugged in, whether or not the receiver is off. Starting an engine, in particular, can induce voltage spikes that can damage the receiver, *even if it is turned off.*

Whether your receiver is powered with regular or rechargeable batteries, when the batteries are almost exhausted the screen fades and the receiver shuts itself down. **In some cases almost no notice is given before the receiver shuts off,** so treat the amount of charge in the battery pack conservatively, as you might the amount of gas in your automobile tank on a long trip, with few fueling stations along the way. For information about the sources of power (battery life, warnings, how to charge, ability to hold a charge while on the shelf, or replacing the batteries) see the receiver operating manual.

{__4} *Different receivers are powered in different ways. Become familiar with yours.*

Juno ST: Find the charging unit. It has an AC plug on one end and a USB Mini plug on the other. (The AC plug is set up so it can accept the variety of international adapters that come with the Juno.) To charge the Juno, plug the "wall wart" into the AC power system, THEN plug the USB connector into the unit. If the battery needs charging a green light will flash on and off on the side of the Juno just above the power switch. When the battery is fully charged the light is a steady green. Turn the unit on and you will see a "plugged in" icon in the upper right of the screen. Touch that and you will get a report on the energy left in the battery. This screen has a lot of settings available. Explore them. One thing to know is that when the receiver is collecting GPS data and putting those data into a file, the backlight will stay on so that you can monitor the process, regardless of the screen power saving settings.

(An environment preserving suggestion: When you are in an air conditioning (cooling) environment unplug the wall wart after the Juno is charged, since it produces a slight amount of heat (a waste) which then has to be moved out of the building by the cooling system (another waste). If you are in an environment where the air is being heated you could leave the charger plugged in, since it (slightly) heats the space, reducing the general heating load of the building. (Purists, you are right: one would have to factor in the relative cost of heating with electricity vs. whatever other method is being used.))

MobileMapper 6: The primary power source for the Magellan unit is two "AA" size batteries. Either disposable alkaline batteries or rechargeable batteries power the unit for several hours. The MM6 can also be run with a "cigarette lighter" plug or an AC power adapter.

Your GPS: Consult the manual. Record the information in your FastFactsFile in your MyGPS folder.

Microsoft Mobile Operating System

It is not reasonable to describe how to use Microsoft Windows Mobile in a book devoted to GPS. So we will go pretty much straight to operating the GPS software—leaving you with the matter of learning Windows Mobile if you don't already know it. (It bears some resemblance to Microsoft Windows but there are major differences as well.) Here are a few vital facts that are important to your use of the handheld as a GPS device:

Things begin with the Start menu. Tap it and then tap Today. Examine the screen.

To see how much battery remains tap Start > Settings > System > Power. Here, under Advanced, you can set the values that let you economize on the amount of time your unit stays on.

To see how much memory remains use Start > Settings > System > Memory.

To set the screen visibility use Start > Settings > System > Backlight (or Brightness).

To look at the folders and files in the receiver go to Start > File Explorer. (Sort of like looking at the C: drive of a PC with Windows Explorer. Instead of C:\ you will find things under My Documents.)

To initiate a program (e.g., TerraSync or ArcPad) go to Start > Programs > program name.

Other Devices and/or Operating Systems

See the various manuals for your system. Record the information in your FastFactsFile in your MyGPS folder.

UNDERSTANDING THE SCREENS AND CONTROLS

Described herein will be two popular GPS software packages for handheld receivers:

- TerraSync from Trimble Navigation
- ArcPad from ESRI

The process of giving commands to a GPS receiver consists primarily of selecting choices from a "Main Menu," from toolbars, and sub-menus.

{__5} *Turn on your receiver.* Learn some basic information about your GPS software.*

---

* The steps that follow have been checked out using a Juno ST GPS receiver. But you might be running some other type of handheld or computer, and may have to make some adjustments.

TerraSync: Tap (or click*) the Start at the top of the Windows Mobile screen. Tap the word TerraSync to start the program. The TerraSync user interface is a combination of sections, subsections, menus, panes, fields, sounds, and graphics. To make matters more complicated, TerraSync can run on a PC (like a laptop) connected to a GPS receiver, or on a mobile device with an embedded GPS receiver, such as the Juno ST. We'll concentrate on the screens you will see on a mobile GPS device, since everything you learn about the handheld transfers without problem to the larger PC monitor.

The highest level of the user interface consists of Sections; there are five of them:

- Map
- Data
- Navigation
- Status
- Setup

The Sections have subsections and options. The whole software structure is based on a hierarchy that starts from the five Sections. To see the entire menu structure search the Internet for

TerraSync "Software Guide"

And, in the 200+ page PDF that you find there, look under "Software Structure."

Figure 1.13 is an image of a TerraSync screen with the Status section prevailing and options for the other sections shown. While we are looking at this screen, which depicts the situation at a particular moment in time, we can note some other information: There are eight GPS satellites (see the number next to the little satellite icon at the top of the screen) being used to compute the position of the receiver,† the battery is fully charged, and the position being computed is about 38 degrees north of the equator and 84.5 degrees west of the prime meridian. The altitude is 280 meters above mean sea level (MSL). The circle with the numbered blocks shows where some of the satellites are—viewed from a position in space toward the antenna. The "N" indicates North. The vertical bar to the right of the circle gives us confidence that there is good reception of the satellite signals. The PDOP (to be discussed later) is 2.42, which also gives us confidence in the accuracy of our position.

---

* TerraSync runs on both handheld devices and Personal Computers (PCs) such as laptops or desktops. When it runs on a PC it runs under the Microsoft Windows operating system. Of course, to collect data that PC must be connected to a separate GPS receiver. In this text we'll assume that you are running TerraSync on a handheld receiver so will use the word "tap" to select an item on the screen.
† Actually it's the position of the antenna, which, with an external antenna, may not be the same as the receiver.

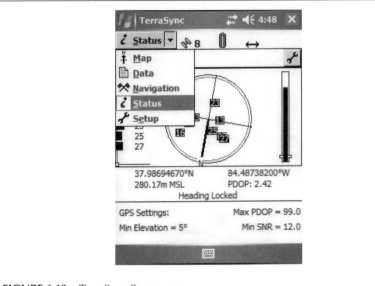

FIGURE 1.13   TerraSync Status screen.

ArcPad: Tap (or click*) the Start at the top of the Windows Mobile screen. Tap the word ArcPad to start the program. The ArcPad user interface is a combination of toolbars with buttons, buttons within buttons, menus, panes, fields, sounds, and graphics. To make matters more complicated, ArcPad can run on a PC (like a laptop) connected to a GPS receiver, or on a mobile device with an embedded GPS receiver, such as the MobileMapper 6.[†] We'll concentrate on the screens you will see on a mobile GPS device, since everything you learn about the handheld transfers without problem to the larger PC monitor.

The highest level of the user interface consists of a Main toolbar (which we'll call Toolbar "A"). See Figure 1.14.

FIGURE 1.14   ArcPad Main toolbar.

---

* ArcPad runs on both handheld devices and Personal Computers (PCs) such as laptops or desktops. When it runs on a PC it runs under the Microsoft Windows operating system. Of course, to collect data that PC must be connected to a separate GPS receiver. In this text we'll assume that you are running ArcPad on a handheld receiver so will use the word "tap" to select an item on the screen.
† Although we are using the Magellan MobileMapper 6 to run the ESRI ArcPad software, it should be emphasized that Magellan makes GPS software that is more native to the MM6, and they also produce PC software into which GPS data can be processed and analyzed. See http://pro.magellangps.com/en/solutions/mobilemap/ for specific information.

Numbering the icons on Toolbar "A" from left to right you have:

1. Open Map Picture (Pic) (1A)
2. Open Map Drop-down Menu (DdM) (2A)
3. Save Map Pic (3A)
4. Add Layer Pic (4A)
5. Add Layer DdM (5A)
6. Table of Contents Pic (6A)
7. GPS Position Window Pic (7A)
8. GPS Position Window DdM (8A)
9. Tools, Toolbars, and Utilities Pic (9A)
10. Tools, Toolbars, and Utilities DdM (10A)
11. Quick Reference Help Pic (11A)
12. Quick Reference Help DdM (12A)

I'll be candid. This is a software user interface for people who know what they are doing. That group of people did not, at least initially, include me. Finding the correct button or menu item can be an exercise in patience. So take good notes and be aware that setting up and using ArcPad is not a trivial matter. Plan to write a lot in your FastFactsFile in MyGPS.

ArcPad was originally meant as a tool for taking pieces of an ArcGIS shapefile into the field for the purpose of updating a database. For example, a utility company may want to periodically check on and record the condition of poles and wires. Facilities exist in ArcPad and ArcGIS for "checking out" a piece of a database, editing it, and checking it back in to make the updates. ArcPad was not designed from the ground up as a GPS data collection device, which is what we are going to use it for initially.

FIGURE 1.15   ArcPad Browse toolbar.

The second toolbar (Toolbar "B") is called Browse. It looks like Figure 1.15.
Unfortunately, the icon names change, depending on what you make active. Here are some mildly informative descriptors:

1. Zoom, Pan, Rotate Pic (1B)
2. Zoom, Pan, Rotate DdM (2B)
3. More Zoom Stuff Pic (3B)
4. More Zoom Stuff DdM (4B)
5. Extents and Bookmarks Pic (5B)
6. Extents and Bookmarks DdM (6B)

7. Identify, Measure, LABEL, and GoTo Pic (7B)
8. Identify, Measure, LABEL, and GoTo DdM (8B)
9. Features—Find and clear Pic (9B)
10. Features—Find and clear DdM (10B)
11. Start/Stop Editing Pic (11B)
12. Refresh Pic (12B)

FIGURE 1.16 ArcPad Edit toolbar.

The Edit toolbar (Toolbar "C"—see Figure 1.16) consists of:

1. Select and Vertex Editing Pic (1C)
2. Select and Vertex Editing Ddm (2C)
3. Feature types Pic (3C)
4. Feature types DdM (4C)
5. Capture Point (from GPS) Pic (5C)
6. Capture Vertex (from GPS) Pic (6C)
7. Capture Vertices (from GPS) Pic (7C)
8. Properties Pic (8C)
9. Properties DdM (9C)
10. Offset point capture Pic (10C)
11. Offset point capture DdM (11C)

FIGURE 1.17 ArcPad Command toolbar.

The Command toolbar (Toolbar "D"—See Figure 1.17) comes up at the bottom of the screen when editing is active.

1. Lock/Unlock the ArcPad Application Pic (1D)
2. Commit Geometry Changes Pic (2D)
3. Proceed to Attribute Capture Pic (3D)
4. Undo Pic (4D)
5. Pen Toggle (to prevent other than GPS input) Pic (5D)
6. Cancel Pic (6D)

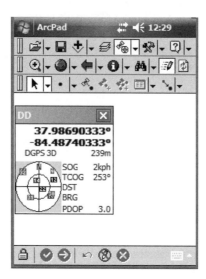

FIGURE 1.18   ArcPad window showing satellite skyplot.

Figure 1.18 is an image of an ArcPad screen with the four toolbars showing and the GPS Position window open. While we are here we can note some other information: There are several GPS satellites being used to compute the position of the receiver,* and the position being computed is about 38 degrees north of the equator and 84.5 degrees west of the prime meridian. The altitude is 239 meters. The circle with the numbered blocks shows where some of the satellites are—viewed from a position in space toward the antenna; the "N" indicates North. Our Speed Over Ground is approximately 2 kilometers per hour. Our course relative to True North is 253 degrees. The PDOP (to be discussed later) is three, which gives us confidence in the accuracy of our position.

Figure 1.19 shows the Signal Chart, indicating the strength of various satellite signals. You bring up this chart by tapping within the circle of the GPS Position Window.

If you tap the circle again you get a compass. See Figure 1.20.

Another tap brings you back to the first window.

---

* Actually it's the position of the antenna, which, with an external antenna, may not be the same as the receiver.

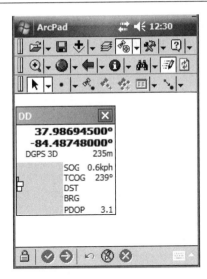

FIGURE 1.19   ArcPad window showing signal strength.

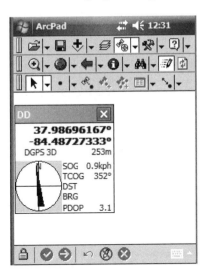

FIGURE 1.20   ArcPad window showing compass.

Your GPS Software: Read over the text in the two previous enclosures. Then use the manuals particular to your receiver to develop similar information in your FastFactsFile.

Setting Vital Parameters

{__6} *Several settings of the GPS receiver must be made correctly; if not, data collection may be hampered or nullified. So set some parameters while you are still inside.*

TerraSync: Proceed as follows: Tap Setup. Tap Reload (to start us off with the defaults the receiver was shipped with). Or, in shorthand: Setup > Reload.* Indicate "Yes." Now: Tap Coordinate System. Choose Bermuda.† Now pick Latitude/Longitude from the drop-down menu. (You can dispense with the keyboard on the screen by tapping the keyboard icon at the bottom of the screen.) Use WGS 1984 for the Datum. Make the Altitude Reference Mean Sea Level (MSL). Use Feet for the Altitude Units. Tap OK.

Tap Units. Set Distance Units to U.S. Survey Feet. Set Velocity Units to Feet per Second. Set Lat/Long Format to decimal degrees (DD.ddd°). The North Reference should be True (toward the North Pole, the axis of Earth's spin) rather than Magnetic (somewhere near 80°N and 115°W; if you stood at the true North Pole the north arrow on your compass would point south). Tap OK.

---

* From now on, when you are to navigate to a certain menu item, I will use this ">" notation. For example "Item1 > Item2 > Item3" means: Tap (or click) Item1, tap Item2, tap Item 3.
† You can dream, can't you?

ArcPad: Start the ArcPad software if it isn't already running. You will (probably) get a screen that asks you how you want to start ArcPad. We won't be using any of the four choices, so dismiss the screen by clicking the white "X" in the red circle.

Activate, if necessary, the GPS: tap the picture of the satellite (7A from the list above), and make the appropriate response. A window named Position should appear. You can drag it down by using the stylus on the title bar so it is out of the way. Tap the adjacent DdM (8A) and you should see red squares indicating that the GPS Position Window is shown and that GPS is Active. Tap a blank spot on the screen to dismiss the DdM.

Since you are not receiving satellite signals yet the unit will begin to complain, using both visual and aural signals. The visual ones get in the way. Let's turn them off. Tap the DdM (8A) next to the satellite picture and pick GPS Preferences. Use the right arrow at the bottom of the screen to find Alerts and tap it. You will see "eye" and "speaker" icons. Clear the checks from the "eye" column and make sure the "speaker" column has all checks. OK.

Tap the Tools picture (9A) and set the Display Units to Statute. Set the Status Bar Coordinate Format to Decimal Degrees (DD). OK the screen.

Tap the latitude and longitude numbers on the square GPS Position Window. Select DD from the menu that appears.

Again go to GPS Preferences (from 8A) and select GPS Height. Make the Antenna Height 1 meter—which is about the height it will be as you carry it around. (The keyboard can be made to appear and disappear by tapping on its icon.) While you are here check that the Datum is set to D_WGS_1984. Under Quality (of the GPS position) make sure that only Non-Compulsory Warnings are issued and that there are checks only next to Maximum PDOP (set the mask value to 8) and 3D Mode Only. OK.

Tap the Tools DdM (10A). You want to clear the red box next to Status Bar if it is on.

A quick checklist before leaving to record positions:

1. Turn on MM6 > Start > Today
2. Check date, fix if necessary
3. Start ArcPad
4. Cancel opening dialogue if necessary
5. Activate GPS Pic (7A)
6. Drag Position box to screen center
7. Set Datum to D_WGS_1984 (8A)
8. Set Display Units to Statute-U.S. Units (9A)
9. Set Maximum PDOP to 8 and use 3D Mode Only (8A)
10. Set Antenna Height to 1 meter (8A)
11. Set all visual Alerts off, aural Alerts on (8A)
12. Turn off Status bar (10A)

The GPS software on your unit has commands and controls described in the manuals that accompany it. Write the relevant information in your FastFactsFile.

Some notes on the above settings:

Depending on your receiver you may be allowed to set the "altitude reference" to either MSL (which is elevation above Mean Sea Level) or HAE (which is Height Above the reference Ellipsoid—the theoretical mathematical surface that approximates the surface of the Earth). Since at this time you probably do not know the relationship between the HAE and MSL at your location, you are selecting MSL.

Depending on your receiver you may be allowed to set masks. In general, a "mask" is a user set value. The receiver compares a given mask with another value that is automatically computed by the receiver. Based on the outcome of the comparison, the receiver uses (or doesn't use) a satellite (or a set of satellites called a "**constellation**") in calculating a position fix. That is, a mask "blinds" the receiver to certain satellites whose signals or positions do not meet the proper criteria for good position finding.

Depending on your receiver you may be allowed to select the "SNR Mask." Each satellite used to compute a position should have a **"Signal-to-Noise Ratio"** (sometimes referred to as signal strength) of four or greater.

Depending on your receiver you may be allowed to set the PDOP mask. We will explain the "PDOP" (Position Dilution Of Precision) term later. For now, just be aware that any PDOP over eight is unacceptable, and four is a figure to use for really precise positioning finding.

Depending on your receiver you may be allowed to set the Antenna Height. One meter (about 39 inches) is a good height since that is about how far you will be holding the antenna over the point whose position you are trying to determine.

Preparing to Correlate GPS Data with Map Data

{__7} *Explore a large scale map of the area you will be taking the receiver to. In the United States a USGS topographic quadrangle (a topo map, usual scale 1:24,000) is appropriate. Note the datum of the map you are using. The datum is usually found in the lower left-hand corner (e.g., NAD-27, the North American Datum of 1927).*

Various distance units you will encounter in work with a GPS include Yards, Meters, Kilometers, Nautical Miles (6,080 feet), Miles (statute, 5,280 feet), "International Feet" (where an inch is defined as 0.0254 meters, exactly), and "U.S. Survey Feet" (where a meter is considered to be 39.37 inches, exactly).

Examine your map to determine the appropriate distance units for the upcoming fieldwork. Feel free to change this value later if you should wish visual output in some other units.

Final Inside Activity

You are almost ready to take the GPS unit into the field. One thing remains to be done. While you are still inside, read through the directions for PROJECT 1B below completely to prepare yourself for the fieldwork. Develop a feel for the sort of data you will be collecting. Practice changing from screen to screen. Outside, with the wind blowing and the traffic roaring, is no time to discover that you don't have a solid surface to write on or that you don't know just what it is you are supposed to be doing. A little preparation now will pay big dividends later.

{__8} *Read over PROJECT 1B below.*

PROJECT 1B

Now Outside

This is an exercise best done with two people. You will take the map, your notebook, and the GPS receiver outside to make observations. You will not yet place the data you collect into a computer file, but you will learn a lot about the factors affecting data collection.

{__1} *As you leave the classroom or laboratory to travel to the site for data collection, be sure the unit is turned on and GPS is activated (see below).*

> **TerraSync:** Tap Setup > GPS and answer the question appropriately.

> **ArcPad:** Tap the picture of the satellite (7A from above), and make the appropriate response.

> See the various manuals for your GPS software. Record the information in your MyGPS > FastFactsFile.

If you carry the receiver exposed to the sky, it will begin to "acquire" satellites. It is <u>not</u> important which menu appears on the display; whenever the receiver is on, it "looks" for satellites and calculates positions if it can.

ArcPad: Turn some visual Alerts back on. See Figure 1.21.

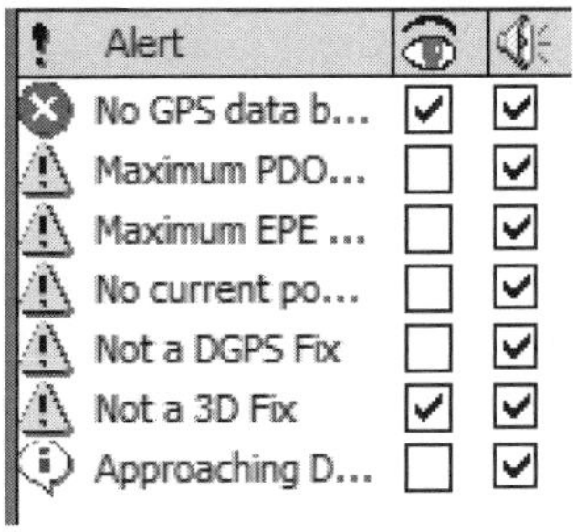

FIGURE 1.21   ArcPad Alerts window.

{__2} *Move to a spot outdoors, well away from buildings and heavy tree canopy. If it is reasonably level and not shrouded by nearby hills or mountains, so much the better. And if you can locate the antenna over a geodetic monument, for which you can find the official latitude and longitude (perhaps from the NGS Web site, http:// www.ngs.noaa.gov), super.*

{__3} *Look at the map to locate your approximate position.*

{__4} *Hold the antenna over the spot for which the coordinates are to be determined. The antenna in a handheld receiver is usually at the top of the receiver, or at the front if you are holding the receiver horizontally. Usually the best way to hold the unit is at about a 45° angle so it is exposed to the sky, you can read it, and your body and head are out of the way of the signals. Hold the unit as far in front of you as is comfortable.*

Actually, no position will be comfortable after a few minutes; you will want to pass the unit to your partner so you can drop your arm and let the blood drain back into your fingers. An alternative is to put the unit on the ground and crouch or sit down so you can read the screen. This is less fun in winter, or when there is poison ivy about. I never claimed fieldwork was easy. You might bring a table or tripod with you, or find a fence post. Be careful: it is easy to bounce the handheld off the ground. It's usually a tough unit but it is also expensive; do you really want to test it?

{__5} *Keep your head and body out of the way, i.e., don't block the signal from a satellite to the receiver. You are opaque, as far as the high-frequency, short-length GPS radio waves are concerned. Remember, the receiver is looking for satellites, some of which are near the horizon. It's easy to forget this and obstruct the antenna, causing the receiver to lose its lock on a satellite.*

{__6} *A few minutes may elapse before the unit locks onto enough satellites to begin giving position fixes. If more than 10 minutes go by with no position fix, restart the software, reactivate the GPS, and make sure you aren't obstructing the signal. You could put the receiver on the ground and move away from it for a few minutes.*

Tracking Satellites

GPS satellites are identified by two-digit numbers. These are the designations, called **PRN numbers**[*] that your GPS receiver uses to identify the satellites. The numbers lie between 1 and 32, inclusive. The numbers that you will see on your receiver are those of the satellites that the receiver might be able to pick up, based on your position and time. They are usually those which are above the horizon and the specified elevation mask angle. The receiver determines which satellites are available by formulas built into its computer and by an **almanac** transmitted by each satellite which describes the general location of all the satellites at that time.

Since you are outside, presumably the receiver is locked onto some satellites. The receiver needs to be receiving at least four satellites before location fixes are computed.[†] Because of geometry relating to the Earth (satellites below the horizon are blocked) the receiver can consider satellites in about one-third of the sky. It may track eight satellites, or even one or two more.

---

[*] "PRN" stands for "Pseudo Random Noise" (no doubt just the designation you expected for the satellites). I will explain later.

[†] If the receiver is in 2D mode, only three satellites are required for a location fix. But unless you have entered the precise altitude of the antenna, the locations calculated by the receiver will be wrong. Don't use 2D mode unless you know what you are doing.

{__7} *Examine how the receiver tracks satellites.*

TerraSync: Select the Status section. Pick Sat Info. Assuming that your unit is receiving satellite signals, the screen will show you the PRN number, the Signal to Noise ratio, the elevation above **your** horizon (an arrow indicates whether the satellite is rising or setting) and the angle from true north that you would have to turn to "see" the satellite. A black dot in the left margin indicates that the satellite is being used for position finding.

Change the subsection Sat Info to Skyplot. Here you get the same sort of satellite information, but shown graphically instead. Black bars show satellites that are used in computing your position. **Also this is the screen that gives you latitude, longitude, and altitude.** If you are using a Juno receiver you will notice that the Max PDOP, Min SNR, and Min Elevation are set and you may recall that you did not set them. That's because the Juno collects almost any signal that reaches it; it is not very discriminating. You will be able to filter out the undesirable points later.

ArcPad: Examine the GPS Position Window that came on when you activated GPS. Make sure positions are being calculated. Change Map Projection to DD (tap the Coordinates to change). The ArcPad manual says that satellites shown in black are used for computation of position, those in blue are available, and those in red are not used. (I cannot verify this; it seems that the unit calculates positions with fewer than four black satellites.)

Your GPS software: you know the drill.

Determining Your Position from the Receiver

TerraSync: Once the receiver is tracking four or more satellites, select the Skyplot subsection, and write down the latitude, longitude, and altitude of the antenna. When locked onto four or more satellites, the receiver computes the position of the antenna about every second. (If the lat/lon/alt numbers begin to flash and/or the message "Too few satellites" appears it indicates that the values presented are those that were collected in the past—perhaps the immediate past—and that the receiver is not calculating new positions. Make certain that there are no obstructions blocking the signals.) If things are working correctly the least significant digits of the position numbers should change about once per second.

> **ArcPad:** The GPS Position Window will show the latitude and longitude. There should be an indication of GPS or DGPS 3D. If things are working correctly the least significant digits of both position numbers should change about once per second.

Your GPS software: Check your FastFactsFile.

{__8} *Note the time. Plan to write down a new position reading in your notebook every minute, approximately on the minute, for the next quarter of an hour.*

{__9} *In between writing position fixes in your notebook you should record some other information. Note down the numbers of the satellites which appear in the window. Circle the numbers of the satellites the receiver is using to compute positions. Also note how many satellites the unit is receiving signals from. Write down the value identified as "PDOP."*

{__10} *Now it is probably about time to write down the next set of position coordinates. They should be close to, but not exactly the same as, those you wrote down a minute ago. The screen should not have flashing values. If it does, you probably got your head in the way of a satellite signal.*

{__11} *Note the bearing or azimuth for each satellite.*

> **TerraSync:** Write down the bearing or azimuth for each satellite. This number specifies the horizontal angle between due north and the satellite: Point your arm toward the north, then rotate your body clockwise until your arm is pointed at the satellite. The number of degrees your body rotated is the bearing or azimuth.

> **ArcPad:** The bearing or azimuth for a satellite is the horizontal angle between due north and the satellite. Look at the GPS Position Window. Pick out a particular satellite. To illustrate the approximate azimuth of that satellite, point your arm toward the north, then rotate your body clockwise until your arm is pointed at where the satellite is said to be. The number of degrees your body rotated is the bearing or azimuth.

See the manuals for your handheld software. Record the information on how to determine the azimuth in your FastFactsFile.

{__12} *Determine the elevations of the satellites being tracked.*

TerraSync: Look again at the screen that shows textual satellite information. You will see several horizontal lines of information—one for each satellite being tracked. One item of information displayed for each satellite is **Elevation**. If you could stand and point a straight arm directly toward the satellite, the elevation would be the angle, in degrees, that your arm made with the Earth, assuming the surface is level where you are standing. Zero degrees would represent a satellite at the horizon; ninety degrees would represent a satellite directly overhead. Write down the elevation of each satellite; it won't change much in the short time you are taking data.

ArcPad: ArcPad does not display satellite elevation information.

Your GPS receiver: See the manuals for your unit. Record the information in your FastFactsFile.

{__13} *Write down another position fix.*

{__14} *For each satellite being tracked, record its signal strength (if your receiver provides this information).*

TerraSync: The last column on the "Sat Posn & SNR" screen is the "signal to noise ratio" (recall that it is an indication of the strength of the signal from the satellite). Acceptable values are greater than or equal to four. Values may range up to 40 or so.

ArcPad: Tap the area of the GPS Position window that shows the locations of the satellites. You will get a bar graph that indicates the strengths of the satellite signals. PRN numbers may or may not be associated with the bars of the graph.

Your GPS receiver: See the manuals for your unit. Record the information in your FastFactsFile.

{__15} *Put your hand over the antenna at the top of the receiver and watch the signal strength drop.*

{__16} *Determine where one or two satellites are in the sky, relative to your position. Try to interpose your body between the unit and a satellite to see if you can make the signal strength drop for a single satellite. In the middle latitudes in the United States there will generally be more satellites to your south than north.*

{__17} *Finish recording the 15 position fixes. Is the unit still tracking the same satellites? Is it using the same constellation of satellites to compute fixes? If not, write down the new information.*

Set Your Watch

I commented earlier that the Global Positioning System might more accurately be called the Global Positioning and Timing System, because it has become the world's time keeper, for most important, practical purposes. The world's time standard was previously called Greenwich Mean Time (GMT). The name for it now is Coordinated Universal Time, abbreviated UTC.* UTC is based on a 24-hour clock.

{__18} *The GPS receiver's clock has been correctly set by the exposure to the satellites. It now has a very accurate idea of the time. So, if your software displays time (ArcPad doesn't) you may set your watch by it and be correct to the nearest second.†* *With the recording of the geographic coordinates and the setting of your watch, you can use GPS to position yourself in four-dimensional (4D) space.*

{__19} *From your receiver (instructions below) determine the current UTC.* _____

TerraSync: Tap the down arrow next to whatever section name appears currently. Tap Status. Tap the down arrow next to whatever subsection appears. Tap UTC Time. Or, in a shorthand version: Status > UTC Time. The UTC time will appear. (The time shown on your screen in the upper right area is the time set in the Windows operating system.)

ArcPad: This software does not indicate time.

Your GPS: See the manuals for your unit. Record the information in your FastFactsFile.

---

* It is called UTC (rather than CUT) because it is based on international agreement and, in French, the adjective "coordinated" comes last.
† The UTC time known by the GPS receiver is correct to a few billionths of a second, but the display is not nearly that accurate.

(As you may or may not know, Einstein's general theory of relativity predicts that time runs more slowly the greater the gravitational field—a somewhat amazing claim that has been verified. That is, a clock would run faster far out in space, away from any large masses, than it would here on Earth and an animal would age more quickly. It turns out that, while the effect is tiny for GPS satellites, their clocks are far enough away from Earth so that they do run at a different rate than those here; and this difference (and another relativistic effect, having to do with time and velocity) has been compensated for.

Did the Earth Move?

If you are standing still why is the receiver indicating that your position is changing?

> **TerraSync:** In the Navigation section tap the Navigate subsection. Set the values of the four boxes at the bottom of the screen to Velocity, Heading, Altitude, and Distance. The Velocity box may show a small value or a question mark. The Heading may also not show a value. The Altitude will keep changing.

> **ArcPad:** Tap the GPS Position circle. Tap it again and you should see an image of a compass. Both the SOG and the TCOG will show some values, indicating that you are moving.

> **Your GPS:** See the manuals for your unit. Record the information in your FastFactsFile.

Any velocity number greater than zero indicates that the antenna is moving with respect to the Earth. That's odd. You see that the antenna is virtually motionless. Why should the receiver be recording movement? The answer, again, is that, with any physical system, there are errors. Your GPS receiver is calculating positions at the rate of about one per second. Since each position differs, slightly, from the one before it, the receiver believes that its antenna is moving.

{__20} *Put the unit in motion, determining as you do so the direction and speed of your course: As you walk with the unit held out in front of you call out the velocity readings to your partner. He or she should "mentally" average your readings and record some values. A fast walking speed is about 6 feet per second or 4 miles per hour.* Your speed indications should be somewhat less.*

---

* A conversion factor that you might want to remember is that 60 miles per hour is exactly 88 feet per second. So approximately two-thirds of a particular foot-per-second number gives you the equivalent miles per hour. Thus, 6 feet per second would be about 4 miles per hour.

{__21} *Continue to walk. Look at the Heading. The course heading number indicates your direction of travel, relative to True North (T), in degrees. Again, call the readings out to your partner. Do they tend to average to the approximate direction you are walking, according to the map?*

{__22} *Walk back to the original location where you recorded the position fixes. Record the latitude, longitude, and altitude. Then shut the unit off and return to your lab, room, or office.*

Project 1C

Back Inside

Your session in the field may have raised as many questions as it answered. We will look at the answers to those questions in later chapters. First, let's verify that GPS really works. (Someone is telling you that you can find your position on Earth to within a few feet from objects in space, more than 12,000 miles away, batting along at 2.4 miles every second. Would you believe them without checking? I wouldn't.)

{__1} *Using a calculator, obtain the average of each of the 15 latitudes, 15 longitudes, and 15 altitudes you recorded.*

Average Latitude _____
Average Longitude _____
Average Altitude _____
Plot the x-y position on your topo map. Does the point represent where you were?* _____ *How about the altitude?* _____

The average altitude indicated developed by GPS is likely to be somewhat different from that shown by the map. The horizontal accuracy of a single point is usually within five to seven meters, or roughly 15 to 22 feet. Vertical accuracy is about half or a third that good. So your altitude fix that your receiver recorded at any given point could be off by more than 40 feet. The average of the 15 altitudes should be somewhat better.

Exercises

(In some cases you can use your receiver to answer the following questions. Or you may need the Internet to work the following.)

---

* You might have to convert decimal degrees to degrees, minutes, and seconds. That's easy to do (with a calculator). Take the fractional part of the coordinate (e.g., 0.87654321 is the fractional part of 43.87654321° and multiply it by 60. This gives you minutes (52.5925926 in this case). To get seconds take the fractional part (0.5925926) of minutes and multiply by 60. You get 35.555555. So 43.87654321° is 43°52'35.55555." Since you are stuck with decimal digits with either process we will ditch the degrees-minutes-and-seconds form wherever practical.

Exercise 1-1

When the GPS system was initiated one of the divisions that was made was on week boundaries, beginning on the midnight between Sunday and Monday. The first week was called GPS Week Zero (0), the next number One (1), and so on. So the GPS week counter starts at zero, then goes up to 1,023 and then resets to zero again. This resetting happened once, in August of 1999. Use either your receiver or the Internet* to find out the current GPS week. What is it? _____ In what year was the original GPS week number zero? _____ Given the GPS week you wrote down above, about how many years would you say have elapsed since the official start of the NAVSTAR GPS project? _____.

Exercise 1-2

When the NAVSTAR GPS was started, GPS time was synchronized with Coordinated Universal Time (UTC). Since that time, however, the spin of the Earth has slowed a bit (due to water sloshing around in the oceans (tidal friction) and other factors). So it has been necessary every couple of years or so to delete so called "leap seconds" from UTC—either between the 30th of June and the 1st of July and/or the 31st of December and the 1st of January.† These seconds have not, however, been used to adjust to GPS time (atomic clocks do not like to be fussed with) so GPS time is now later than UTC time. GPS time is what Earth time would be if they hadn't taken out some seconds that occurred in Junes and Decembers. Use the Internet or other source to determine how many seconds GPS time is ahead of UTC time. _____

---

* You might try http://www.csgnetwork.com/timegpsdispcalc.html, which might be found by searching the Internet for "Multiple Time Display Converter."

† "Leap second" is something of a misnomer, if one compares it with "leap year." February 29 is a "leap day" which occurs in a "leap year." But a leap second performs the same purpose as a leap day: it delays the restart of the regular time keeping system by some amount.

GPS Equipment Check-Out Form

We (please print names, telephone numbers, e-mail addresses)

_____ , _____ , _____

_____ , _____ , _____

wish to check out a GPS receiver:

Handheld Receiver Type _____

on _____ ____/____/____ at _____ _____ .
    (day)   (date)   (time)   (place)

We understand that this equipment is valuable, and we will use our best effort to guard it against damage, loss, or theft.

We intend to take data at _____

which is at approximately latitude _____

and longitude _____ .

We will return said equipment

on _____ ____/____/____ at _____ _____ .
    (day)   (date)   (time)   (place)

_____   _____
(signed)   (signed)

In case of difficulty in returning the equipment we will contact _____ , at phone number _____ . We will leave a message on the recorder if there is no answer.

_____

Office Use:

Unit # _____ checked *OUT* at _____ on _____ by _____

Unit # _____ checked *IN*   at _____ on _____ by _____

2 Automated Data Collection

In which you learn the basic theoretical framework of GPS position finding, and practice using a GPS receiver to collect computer-readable data.

OVERVIEW

HOW'D THEY DO THAT?

By now you have determined that GPS really works. That little gadget can actually tell you where you are! How?!

The fundamentals of the system are not hard to comprehend. But misconceptions abound, and it is amazing how many people don't understand the principles. In a few minutes, assuming you keep reading, you will not be among them.

We look first at a two-dimensional analogy. You are the captain of an oceangoing ship off the western coast of some body of land. You wish to know your position. You have aboard

- an accurate timepiece;
- the ability to pick up distant sound signals (a megaphone with the narrow end at your ear, perhaps?);
- a map showing the coast and the locations of any soundhouses (a sound-house is like a lighthouse, but it emits noise instead); and
- the knowledge that sound travels about 750 miles per hour, which is about 20 kilometers (km) per minute, or 1/3 km per second.

Suppose:

1. There is a soundhouse located at "S1" on the diagram in Figure 2.1.
2. Each minute, exactly on the minute, the soundhouse horn emits a blast.

With these elements, you can determine the distance of your vessel, V, from the soundhouse. To do so, you note when your clock marks an exact minute. Then you listen for the sound signal. When it comes, you again note the second hand of the clock. You then may calculate d from

$$d = s/3$$

where d is the distance in kilometers from the soundhouse and s is the number of seconds it took for the sound to reach you.

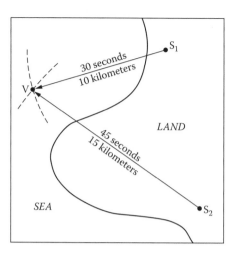

FIGURE 2.1   Measuring distance by measuring time.

Suppose it took 30 seconds for the signal to arrive. You would know your ship was 10 km from soundhouse S1.

In geometric terms, you know only that your ship is located somewhere on the surface of a sphere that has a radius of 10 km. You can reduce this uncertainty considerably because your ship is floating on the ocean, so you know your altitude is fairly close to mean sea level (MSL). Thus you could consider yourself to be on a circle with a radius of 10 km.

That, however, does not pinpoint your location. And, by looking at the diagram, you can see that, if you are moving, you might be in a lot of trouble, because contact with the ground is not recommended for ships. How could you determine your position more exactly? By finding your distance from a second soundhouse.

Suppose you use the same technique as above, listening to soundhouse "S2." You find it is 15 km away. Then the question is, where is (are) the point(s) that are 10 km away from "S1" and 15 km away from "S2"?

If you solve this problem graphically, by drawing two circles, you find they intersect in two places. One of these locations you can pretty well eliminate by noting that your ship is not sitting on a prairie or in a forest. Based on the measurement from the second soundhouse, you know your position as accurately as your measuring devices and map will let you know it.

It is obvious that you can find your position by drawing circles. You can also find your position by purely mathematical means. (You would determine the coordinates of the intersection of the two circles from their formulas by solving two "simultaneous" equations.) Then you would consider lengths of the sides of the triangle formed by the two soundhouses and the ship. This process is sometimes called *trilateration*.

As with the soundhouses, distances—determined by the measurement of time—form the foundation of GPS. (Such distances are referred to as ranges.) Recall the basis for the last three letters of the acronym NAVSTAR: Time And Ranging.

How It Works: Measuring Distance by Measuring Time

How do the concepts illustrated above allow us to know our position on or near the Earth's surface? The length of that answer can vary from fairly short and simple to far more complicated than you are (or I am) interested in. At its most detailed GPS **is** rocket science, brain surgery, nuclear physics, and the theory of relativity all rolled into one.

Basically, distances to several satellites (four are needed for a good 3D spatial fix) are calculated from measurements of the times it takes for radio waves to reach from the satellites, whose positions are known precisely, to the receiver antenna.

To illustrate, let's look at these basic ideas by seeing how the "ship" example differs from NAVSTAR.

- NAVSTAR gives us 3D locations: Unless we are in fact on the sea, in which case we know our altitude, the problem of finding our location is three-dimensional. GPS can provide our position on or above the Earth's surface. ("Below" is tricky because of the radio wave line-of-sight requirement.) But the method translates from two to three dimensions beautifully.
- The "soundhouses" are satellites: Rather than being situated in concrete on a coast, the device that emits the signal is a satellite, zipping along in space, in an almost exactly circular orbit, at more than 2 miles per second. It is important to note, however, that at any given instant the satellites (space vehicles, or SVs) are each at one particular location.
- NAVSTAR uses radio waves instead of sound: the waves that are used to measure the distance are electromagnetic radiation (EM). They move faster than sound—a lot faster. Regardless of frequency, in a vacuum EM moves at about 299,792.5 km/s, which is roughly 186,282 statute miles per second.

Let's now look at the configuration of GPS with a drawing of true but extremely small scale, starting with two dimensions. Suppose we represent the Earth not as a sphere but as a disk (like a coin), with a radius of approximately 1 unit. (One unit represents about 4,000 statute miles.) We are living on the edge (pun intended) so we are interested in points on, or just outside (e.g., airplanes and orbiting spacecraft) the edge of the coin. We indicate one such point by x on the diagram of Figure 2.2. We want to find out where x is.

Suppose now that we have two points on the drawing of Figure 2.3, called a and b. They represent two of the NAVSTAR satellites, such that we can draw straight lines from a to x and from b to x. We require that the lines not pass through our "Earth." These points a and b are each about 4 units from the center of the coin (thus 3 units from the edge of the coin).

Measure the distance, or length, from a to x (call it La) and from b to x (call it Lb). The lines La and Lb represent the unobstructed lines of sight from the satellites to our receiver antenna. If we know the positions of a and b and the lengths of La and

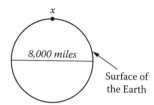

FIGURE 2.2 The Earth and a point to be found.

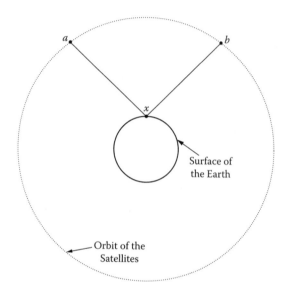

FIGURE 2.3 The Earth and GPS satellites.

Lb, we can calculate the position of *x* through the mathematical process of trilateration* shown in Figure 2.4.

It is intuitively obvious that we can locate any *x*, given knowledge of the positions of *a* and *b*, and of the lengths *La* and *Lb*: *x* must lie on a circle centered at *a* with radius *La* and also must lie on another circle centered at *b* with radius *Lb*.

To formalize a bit what we learned from the example with the ship and the soundhouses, two such circles either (1) do not intersect, (2) touch at a single point, (3) touch everywhere, or (4) intersect at two points. Possibility (1) is clearly out, because we have constructed our diagram so that the circles intersect at *x*. Possibility (2) seems unlikely, given the geometry of the situation, and given that the line segments connecting the points with *x* must not pass through the disk. Possibility (3) is out: the circles would have to have the same centers and radii. We are left then with possibility (4), shown as Figure 2.5: there are two points that are at distance *La* from *a*

---

* A frequent mistake is to say that GPS operates by triangulation. Trilateration measures the distances related to a triangle; triangulation measures angles. GPS uses trilateration. A triangle is a trilateral, just as a four-sided figure, such as a rectangle, is a quadrilateral.

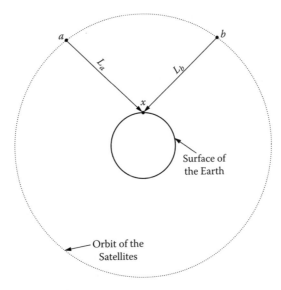

FIGURE 2.4 Distances from x to the GPS satellites.

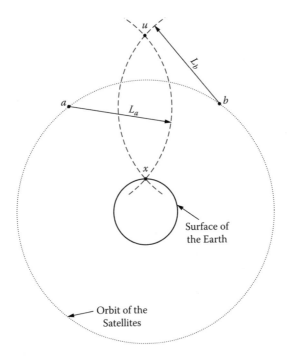

FIGURE 2.5 Finding x by the intersection of circles.

and *Lb* from *b*. One of these is *x*. The other is a long way from the disk, is therefore uninteresting (we'll call it point *u*), and is certainly not *x*.

The points *a* and *b*, of course, represent two of the GPS satellites at a precise instant of time, and point *x* represents the position of the GPS receiver antenna at that same precise moment. (The 0.1 second that it takes for the signals from the satellites to reach the receiver in compensated for in the design of the system.)

This is fine for two dimensions. What happens when we move to three? From a conceptual point of view, the coin becomes a sphere, *x* resides on the sphere (or very slightly outside it), and we need to add a point (satellite) *c*, somewhere else in space (not in the plane formed by *a*, *b*, and *x*) in order to be able to locate the position of *x*. The problem we solve here is finding the intersection of three spheres (instead of two circles).

Although the process of finding this intersection is more difficult mathematically, and harder to visualize, it turns out, again, that there are only two points, just one of which might reasonably be *x*.

You may deduce this as follows. Consider initially the intersection of the surfaces of two spheres. First, forget about the special case possibility that they don't intersect at all (their centers are separated by more than the sum of their radii, or one sphere completely contains the other), and the case in which they touch everywhere (same centers, same radii). Now they must either touch at a single point (unlikely) or intersect in a circle. So our problem of visualization is reduced to finding the intersection of this circle and the surface of the third sphere. If you again discard some special, inapplicable cases, you see that the intersection can be only two points—only one of which can be *x*.

In theory, one should need only three satellites to get a good, three-dimensional (3D) fix. You may recall that when you took the receiver into the field, however, you needed four satellites before the unit would calculate a position. Why? Briefly, the reason has to do with the fact that the clock in a receiver is not nearly as good as the four $50,000 clocks in each satellite, so the receiver must depend on the satellite clocks to set itself correctly. So, in a sense, the fourth satellite sets the receiver's clock. In actuality, however, all four satellites contribute to finding a point in four-dimensional space.

You now know the theoretical basis for GPS. It is not unlikely that you have several questions. I will defer the answers to a few that we have anticipated so that we can take up an issue that bears more directly on the fieldwork you are about to do.

FACTORS AFFECTING WHEN AND HOW TO COLLECT DATA

Neither GPS nor any other method can tell you exactly where an object on the Earth's surface is. For one thing, an object, no matter how small, cannot be considered to be in exactly one place, if by "place" we mean a zero-dimensional point specified by coordinates. Any object occupies an infinite number of zero-dimensional points. Further, there are always errors in any measuring system or device.

Now that we have thrown out the idea of "exact" location, the issue becomes: what kind of approximation are you willing to accept (and pay for)? There are two philosophies you might consider:

- Good enough
- The best that is reasonable

You might use "good enough" when you know for certain what "good enough" is. For example, if you are bringing a ship into a harbor, "good enough" might be locational values guaranteed to keep your ship out of contact with the harbor bottom, or between buoys.

You might, on the other hand, use "the best that is reasonable" when you don't really know what "good enough" is. For example, if you are building a database of city block outlines, you might not be able to foretell the uses to which it might be put. Your immediate needs might suggest that one level of accuracy would be appropriate, but several months later you (or someone else) might want a higher level of accuracy for another use. So it might be worth expending the extra resources to collect data at the highest level of accuracy that your budget and the state of the art will allow.

The major factors that relate to the accuracy of GPS measurements are:

- Satellite clock errors
- Ephemeris errors (satellite position errors)
- Receiver errors
- Ionosphere errors (upper atmosphere errors)
- Troposphere errors (lower atmosphere errors)
- Multipath errors (errors from bounced signals)

There are several fifty-cent words above that haven't been defined. We will defer discussion of most of the sources of error until later, primarily because, at the moment, there is little or nothing you can do about them besides knowing of their existence and understanding how they affect your results. In the long run you can do a lot about these errors by a method called "differential correction." But first we will consider an issue called "Dilution of Precision" (DOP) because:

- High (poor) DOP values can magnify the other errors,
- DOP values can be monitored during data collection and the data logger can mask out data with excessive DOP values,
- DOP values, which can be predicted considerably in advance for any given location, can perhaps be reduced by selecting appropriate times to collect data, and
- Differential correction cannot eradicate errors created by inappropriate DOP values.

Position Accuracy and DOP

Prior to the end of 1993, GPS had less than the full complement of 24 satellites operating. In earlier years, there were periods during the day when there were not enough satellites in view from a particular point on the ground to provide a position fix.

Now a data collector can almost always "see" enough satellites to get a position fix. But the quality (accuracy) of that fix is dependent on a number of factors, including,

- the number of satellites in view, and
- their geometry, or arrangement, in the sky.

In general, the more satellites in view, the better the accuracy of the calculated position. The receiver you are using, and many others, may use the "best" set of four satellites to calculate a given point—where "best" is based on a DOP value. With more satellites in view, there are more combinations of four to be considered in the contest for "best." Data collection with fewer than five satellites in view is pretty iffy and should be avoided when possible. I'll say more later (in Chapter 7) on how you can know ahead of time when enough satellites are available and what the DOP values will be. It is always possible to see four (with 30 up and broadcasting properly—"healthy," as they say) and you might occasionally see more than twice that number.

If more than four satellites are available "over-determined position finding" can be used. This means a receiver can use more than four satellites to determine a position. Since a set of four can determines a position, a set of five could determine five positions—by considering satellites 1234, 1235, 1245, 1345, and 2345. The "average" of these positions would probably be better than any one alone, but a more complex statistical method is used to provide the most accurate position. The statistical treatment of the use of more than four satellites is a complicated matter and will not be addressed further. Presumably over-determined position finding results in a more accurate result.

The concept of DOP involves the positions of the satellites in the sky at the time a given position on the ground is sought. To see why satellite geometry makes a difference, look at the two-dimensional case again shown by Figure 2.6. Suppose first there are two satellites (*a* and *b*) that are being used to calculate a position *x*: If we know the distances *La* and *Lb* exactly, we can exactly find the point *x*.

But we don't know the *La* or *Lb* exactly, because of the error sources listed above. For illustration, suppose that we have an error distance, delta, that must be added and subtracted from each of *La* and *Lb*. That is, for each distance there is a range of uncertainty in the distance that amounts to two times delta. This is illustrated graphically by Figure 2.7.

Now suppose there are two satellites (*c* and *d*) which are being used to calculate a position *y*. These satellites are further apart than *a* and *b*, so that the angle their lines make at the receiver (in Figure 2.8) are more obtuse than the almost-90 degree angle from *a* and *b*.

If we know the distances *Lc* and *Ld* exactly, we can exactly find the point *y*. But again we do not know the lengths exactly—we use the same difference of "two times delta" (in Figure 2.9).

The shaded figure shows the range of positions the receiver might indicate for fix *y*. As you can see, the distance between the true position and the position that could be reported by the receiver for the second case is considerably larger than for the first case.

My goal here is to illustrate that "**satellite geometry**" can make a big difference in the quality of the position you calculate. In actual GPS measurements, of course, it is a volume rather than an area that surrounds the point being sought, but the same general principles apply.

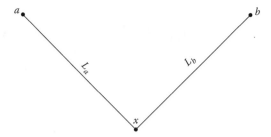

FIGURE 2.6 Satellite positions relative to x.

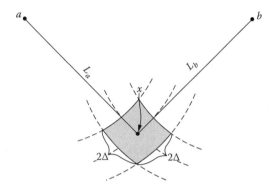

FIGURE 2.7 Area in which x might reside, given satellite positions a and b. The shaded figure indicates the area that contains the actual location sought.

FIGURE 2.8 Satellite positions relative to y.

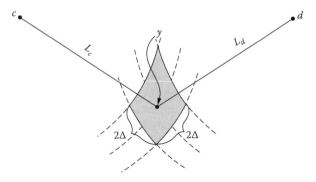

FIGURE 2.9 Larger area of uncertainty due to satellite positions.

So, Actually, What Is DOP?

DOP—sometimes referred to as GDOP ("Geometric" Dilution of Precision)—is a number which is a measure of the quality you might expect from a position measurement of the GPS system based solely on the geometric arrangement of the satellites and the receiver being used for the measurement. If you think in terms of geometric strength you can pretend the electronic ranges from the satellites are cables and you want the strongest arrangement you can have to keep the receiver antenna from moving much when the cables stretch. Ideally you would like to have the four satellites at the vertices of a regular tetrahedron (a three-sided pyramid with all four surfaces equilateral triangles) and the receiver antenna at the center. This would provide the greatest geometric strength, but of course it is impossible for GPS, because three of the satellites would be below the horizon. The best alternative would be when one satellite is directly overhead and the three others are close to the horizon, spaced on a circle, separated from each other by 120 degrees. Simply put, you get the best DOP when the satellites are spread out; the more satellites clump together the worse the DOP value. In summary, DOP is a measure of the extent to which satellite geometry exacerbates the other errors that may occur in the measurement.

The overall DOP number is made up of several "sub-DOPs":

- HDOP (Horizontal DOP) is a combination of NDOP (North DOP) and EDOP (East DOP),
- VDOP is Vertical DOP,
- PDOP (Position DOP—generally considered the best single indicator of geometric strength) is a combination of HDOP and VDOP (actually the square root of the sum of squares of HDOP and VDOP),
- TDOP is Time DOP, and
- GDOP ("Geometric" DOP) is a combination of PDOP and TDOP.

You may recall that (depending on which receiver you used) when you set up the GPS receiver in Chapter 1 you may have set the maximum allowable PDOP. PDOP is the most important single DOP to consider. As you saw, on one screen you set the PDOP value; on another screen, during data collection, you got a report on the value of PDOP of the constellation the receiver was using. The PDOP mask says: "If the PDOP is too high then don't calculate any positions." The recommended PDOP mask value settings are 1 to 4 (great), 5 or 6 (okay), 7 or 8 (marginal), and greater than 8 (unacceptable).

STEP-BY-STEP

During the projects in Chapter 1 you visually read and manually recorded, with pencil and paper, positions that were calculated by the GPS receiver. In this session, using the same general geographic locations as last time, you will use the memory capacity of the receiver to store the readings in machine-readable (i.e., computer-readable) format. That will be Project 2A. Later, in Project 2B, you will also take data as you move the antenna along a path by walking, bicycling, or automobile. During both data collection sessions the position data will be automatically collected into computer files. Project 2C will ask you to upload the collected data into a PC.

PROJECT 2A—RECORDING SINGLE POINTS IN THE DATALOGGER'S MEMORY

Inside: Planning the GPS Data Collection Session

For the projects in this chapter you will need your notebook, the receiver, and, optionally, the external antenna, if you have it. Set aside a section of the notebook to record information about the files you will collect. On a form, such as the one found at the end of this chapter, you should plan to manually record:*

- the date and time,
- a general description of the location or path that is the subject of your data recording,
- the starting time and the ending time of the data collection,
- the filename,
- the amount of memory available in the datalogger before data collection, and then, when finished collecting a file, the actual size of the file,
- the interval between collected data points (e.g., every 10 seconds, or every 50 meters),
- the values of PDOP,
- the number of satellites being tracked,
- the hours of charge remaining in the battery, and the amount of time it is used during this session.

You are going to do some hand calculating once you get back inside. The most important lesson you should take from this book is that when you combine spatial data you have to be sure that the data sets agree in datum, coordinate system, projection, and units. I plan to drive that point home at the end of the chapter, but you need to capture a point while you are in the field so I can illustrate it. So you will have to put up with a bit of "clickology" (pushing buttons without really knowing why) to capture a geographic point—sometimes called a "way point."

Setting Up the Receiver/Datalogger†

Your goal here will be to ensure that the data you collect during your trip to the field will be worth something when you get back.

The receiver obtains data from the satellites and calculates positions at about one per second, but the datalogger is usually set by the user to record point fixes in the microcomputer's memory less frequently. You have some control over how often point fixes are recorded. The datalogger also records the exact time each point fix is taken—a fact whose importance will become apparent later.

{__} *Be sure that the battery pack is sufficiently charged to complete the session with an ample reserve. Use the operating system of the handheld or laptop to indi-*

---

* Not all receivers will allow you to know all of the listed items.
† For purposes of this discussion, we will speak of the "receiver" as a device that obtains the fixes from the satellites, and a "datalogger" as a device that records these data into a file. The equipment you are using houses the receiver and datalogger in the same physical unit.

cate how much charge remains in the battery. In terms of % it is approximately _____ . *When in doubt, recharge.*

{__2} *Be sure that you have sufficient memory to do a project. If less than 20% of the memory is unused you should try to clear away some old files. Again, use the operating system of the unit.*

{__3} *Ensure that the receiver settings are proper. Look in Chapter 1 and set up the receiver according to the instructions there.*

IN THE FIELD: COLLECTING DATA

Take the receiver outside. You are going to collect two data sets consisting of position information. The first will consist of one or more single points. The second will be of a sequence of vertices which will represent a linear feature such as a stream or road. You will deal with collecting unique named points first. With TerraSync you will make what is called a Waypoint. If you are using ArcPad you will, after some setup, collect three named points.

TerraSync: Go to the Navigation section. You will be creating two things: (1) A file that can contain a number of waypoints, and (2) a particular waypoint inside that file. In the subsection menu tap Waypoints. Tap New to create a "GPS Waypoint File." Use the miniature QWERTY keyboard to name the file GPSWPF_yis, where yis are your initials. (My file, for example, would be GPSWPF_MK.) This will create a file that will contain the particular positions you designate later. OK.

To create an actual point Tap Options > New. Place SinglePoint in the Name Field. Tap Create From and pick GPS, which will fill in the Latitude, Longitude, and Altitude fields. OK. (You will see that SinglePoint has been recorded as point #1 in waypoint file GPSWPF_yis. Wait a few seconds and then look at the Distance field. It will show the number of feet the receiver antenna is from the waypoint. Why isn't the distance zero, since you just made the point from the GPS position and you haven't moved? For the same reason the 15 readings you took in the last chapter are all different: error.

ArcPad: In addition to the settings from Chapter 1 you will need to set parameters for recording data in files. You may recall that ArcPad was originally designed for taking a portion of an already established data base into the field, editing it, and returning the edited data back to the data base. It was not originally designed primarily as a data collection system. But the ability to collect data "from scratch" has been added; it is named QuickProject and it appeared in ArcPad Version 7.1. It allows the software to work with files that are generated on the spot, rather than being "checked out of" a database. We will use QuickProject in this chapter.

We are going to collect two folders of information. The first will consist of single points. The second will be of a sequence of vertices which will represent a linear feature such as a stream or road. Each of these will be named "QuickProject YYYY-MM-DD HHXX." Where YYYY is the year, MM the month, DD the day, HH the hour, and XX the minute the folder was created.

First be certain that GPS is active. Next Enable Averaging under Capture under GPS Preferences, found under the GPS DdM (8A). Specify 30 Points to be averaged. You will be "editing" a blank data set, so you will need the Edit toolbar. Bring that up with the Tools DdM (10A). The Edit toolbar appears, with every icon grayed out—because there is nothing to edit yet. You will also notice that the Command toolbar also emerges at the bottom of the window.

To initiate the QuickProject you go to the Open Map DdM (2A) and tap New. Tap QuickProject. Read and OK the screen. If asked whether to save changes to Untitled say "No." If necessary uncheck "Keep layers from current map." OK. Write down the Project folder name, so you will be able to distinguish it from others in the datalogger. Notice that you have been provided three shape files: Points.shp, Lines.shp, and Polygons.shp. OK.

You are returned to the main window. Notice that parts of the Edit toolbar have come alive.

Tap the pencil—the Start/Stop Editing Pic (11B)—and you will notice that all three feature types are available for editing. You can only edit one at a time so turn off the red boxes around Lines and Polygons.

Tap the Feature types DdM (4C) and select Point. Tap the icon Capture Point from GPS Pic (5C) and you will be presented with a screen asking for a Name for the point. Call it SinglePoint. While you are putting in the name of the point you will notice that at the top of the screen is a counter that gives an increasing percentage. This indicates progress toward collecting the 30 points you asked to be averaged. When the counter reaches 100% and disappears, tap OK. In the same way make a point named Point #2 and another: Point #3 from GPS.

You will notice that you only see the label SinglePoint on the map. That's because the points are very close together and, as you can see from the scale bar, a lot of real estate is being shown. So use the Zoom In Pic tool (2B) and repeatedly drag a box around Point #1. Soon you will see the other points. If you zoom in too far, use the Previous Extent Pic (5B) or the Zoom Out tool (found on the Extents DdM) to back out some. You can also Zoom to Full Extent with 4B DdM. Observe the Scale Bar.

Ideally, all these points would lie in exactly the same place, since the antenna hasn't moved. But there are errors and these result in different positions for each of the points. The red circle with + in it is the current position that the GPS is calculating. ArcPad will attempt to keep it on the screen, so the points may be moved out of view if the calculated position varies too much from the locations of the three points. You can prevent the window from always keeping the current position on the map by going to GPS Preferences (you remember where that is by this time) and clicking off Automatically Pan View.

You have completed the process of taking points from a fixed location. If things didn't go according to plan, try again, using the summary below.

Summary of Hints for Collecting Points:

1. Turn on MM > Start > Today
2. Check date, fix if necessary
3. Start ArcPad
4. Cancel opening dialogue if necessary
5. Activate GPS Pic (7A)
6. Drag Position box to screen center
7. Set Datum to D_WGS_1984 (8A)
8. Set Display Units to Statute-U.S. Units (9A)
9. Enable Point Averaging (8A)
10. Set Maximum PDOP to 8 and use 3D Mode Only (8A)
11. Set Antenna Height to 1 meter (8A)
12. Set all visual Alerts off, aural Alerts on (8A)
13. Turn off Status bar (10A)
14. Make sure positions are being calculated
15. Change Map Projection to DD (Tap Coordinates to change)
16. Display Edit toolbar (10A)
17. Start QuickProject (no current layers)(2A > New)
18. Enable Points only with Pencil Pic (11B)
19. Enable Point with Feature DdM (4C)
20. Capture Point from GPS Pic (7C)
21. Name the Point—wait for collection to finish
22. Capture next point
23. Name the Point—wait for collection to finish
24. Capture third point
25. Name the Point—wait for collection to finish
26. Start a new QuickProject **OR**
27. Exit ArcPad with Open Map DdM (2A)

See the various manuals for your software. Record the information in your FastFactsFile. Collect three GPS fixes.

By the way, your datalogger has the capability of recording **attributes** of point features (e.g., the number of a parking meter), but we won't be using that ability until the exercises in Chapter 6.

Project 2B—Taking Data along a Path

In this project you will again use the receiver to take fixes and record the coordinates in the datalogger. The difference between this and the previous project (2A) is that you will move the receiver antenna through space in order to record data points along a track. This capability allows you to generate the more interesting features of a GIS: lines and polygons, rather than simply points.

You may select one of three ways to move the antenna: by foot, bicycle, or automobile. Read the sections below to decide how you want to take data. (In the event that you cannot take moving data, several data files are provided on the CD-ROM that accompanies this book, so you may process those files. Files found there include those taken by automobile, bicycle, airplane, helicopter, sailplane, cruise ship, and sailboat. I have no data presently from a hot air balloon; if you generate some, please feel free to send it in.)

Generating a series of points while moving provides some challenges:

- Accuracy. We indicated that 180 fixes at a single location provided a reasonable approximation of the position of the antenna. In collecting data while moving, of course, you will collect only one fix at each location, with the concomitant loss of accuracy.

- Constellation Vacillation. When you set up the receiver at a single point, you can try to optimize the view of the satellites by staying away from obstructions. While moving, you have little opportunity to pick the points at which the receiver calculates a position. As the antenna moves along the path to different positions, the receiver, by simply trying to pick the best set of satellites to calculate positions, may choose different constellations of satellites, due to signal obstruction caused by tree canopy, buildings, overpasses, or tunnels. Because of the various errors, which are different from satellite to satellite, a position fix reported by the receiver using a new constellation may be different from where it would have been had the previous constellation been retained. Thus, the position fixes may not follow a clean line, but may jump from side to side of the true path. A second consideration in taking data along a path is that each change in constellation increases the amount of memory necessary to store data.

- Multipath. Substantial errors may occur if a given radio signal follows two paths to the receiver antenna. This can happen if a part of the signal is bounced off an object, such as a building. The arrival of two or more parts of the signal at different times can confuse the receiver and produce a false reading. (You may have seen the results of multipath on the screen of a television set receiving signals from an antenna (not from cable); it appears as "ghosting"—a doubling or tripling of an image. Many GPS receivers are programmed to disregard the second signal. But a problem can occur if the direct signal is blocked but the related bounced signal is seen by the antenna and recorded.

TAKING DATA ON FOOT

While walking, you need to be careful to keep the antenna high enough so that no part of your body impedes the signal. With a receiver with a built-in antenna, this means holding it high, well out in front of you, so that the internal antenna can have a clear "view" of the sky. With a receiver with the external antenna, it is probably also a good idea not to wave the antenna around any more than is absolutely necessary. A pole, with the unit or antenna affixed to it, attached to a backpack is a nice solution.

When walking, you might want to think about setting the interval between logged points based on time, not distance, although you might think that taking a point every so many feet or meters would seem like a good idea. The problem is that the receiver will record almost all spurious points that occur, because it is set to record a point that is more than "d" distance away from the last point. If "d" is, say, set to 20 feet, and errors create a point 25 feet away from the last point, this spurious point will be recorded. (Recall that the receiver actually generates points at about one per second.) If, instead, you are recording points, say, every 4 seconds, then there is a better chance that a spurious point may be ignored. You may certainly try this both ways, but you will probably get a ragged path with many spikes in it by using distance as the logging interval while walking.

A brisk walking speed is about 5 feet per second, so you can set the time interval accordingly, depending on the point spacing you want. For this project, if you are walking, select a path, preferably closed, of a mile or two.

COLLECTING DATA BY BICYCLE

An economical way to collect data over significant distances along a linear path is to use a bicycle. But there are a number of pitfalls, in addition to the general dangers all bicyclists face, mostly from automobiles. The principal problem relates to the line-of-sight requirement: unless the antenna is positioned far away from your body, or above it, the receiver will consistently lose its lock on one satellite or another.

One solution—not recommended—is to attach the antenna to the highest point on the "you+bicycle" combination, but bicycle helmets are pretty nerdy-looking appliances anyway; when you duct tape a GPS antenna to the top of it the effect is, well, startling—to the point that neighbors come out with cameras to capture the image. One solution is to "shoot the moon" as it were, to give up any pretense of self-respect, and clothe the antenna in one of those "propeller beanies."

Some other solutions:

- have a special bracket made for a surveying pole with the antenna affixed,
- wear a backpack containing the receiver with the antenna on a pole,
- ride a tandem bike (a bicycle built for two), and attach the antenna to the second seat. The effect is still pretty ridiculous but let's face it: you aren't going to collect GPS data on a bicycle while maintaining a significant level of dignity anyway.

COLLECTING DATA BY AUTOMOBILE

To collect data by automobile you should ideally use the remote antenna with a magnetic mount. Place the antenna on top of the car and run the antenna lead through a window, preferably in a rear door. Some safety tips:*

---

* The reader could speculate on how the author can be so detailed in his description of the problems that might arise, but the time would be better spent learning about GPS.

- Put a thin pad of cloth between the magnetic mount and the car roof to prevent scratching the car's finish.
- If you want to roll up the car's window be careful not to crimp the antenna wire between the glass and the frame; this is especially easy to do with power windows.
- Do not open the door through which the antenna lead is threaded.
- Leave enough slack in the antenna lead outside the window so that when you *do* accidentally open the "antenna lead" door you don't drag the antenna across the roof.

If you are not using a remote antenna, place the receiver as far forward on the dashboard, under the windshield, as you can, so that metal of the car's roof blocks signals as little as possible.

One advantage to collecting data by car is that you can use the power supply from the vehicle. *However, remember not to start or stop the engine with the receiver attached to the car's power supply.* The proper order of events is as follows:

1. Start the car engine.
2. Connect the plug to the car's auxiliary power (cigarette lighter) receptacle.
3. Turn on the receiver.
4. When you are ready, start recording data.

When you are through, **undo** the steps above, *in **reverse** order.*

With automobile data collection, you may use either time or distance as the logging interval. While errors may still be a problem, the distance you set between logged points is going to be much greater with the car than while walking. Thus, the chance of a point being generated by a multipath event that is greater than the logging interval is reduced considerably.

While riding as a ***passenger*** in the auto (the driver is supposed to keep eyes on the road) examine the navigation screen occasionally. You will notice that it gives the car's speed within a mile per hour or so, and the direction as well. (You may set the units of display so as to get miles per hour; changing units will not affect data collection, although pulling the receiver back from its place at the front of the dash may.)

ACTUAL DATA COLLECTION

{__1} *Fill out the initial portions of the Data Collection Parameter Sheet. Choose a method of transporting the antenna. Choose a route—perhaps a closed loop. Decide on a recording interval.*

{__2} *Start the data recording process, using the appropriate procedure below.*

TerraSync: First be certain that TerraSync is running and GPS is active. Set the Units and Coordinate System as before. Next you may want to specify that a vertex is taken only when the antenna is moved a certain distance. You'll find this under Setup > Logging Settings > Between Feature Logging. Set the Style to Distance. If you are walking with the receiver maybe you want to specify the Interval as 10 feet. If on bicycle perhaps 30 feet. In an automobile you could use 200 feet. OK. Also under Logging Settings set Log Velocity Data to Yes. Assure that the Filename Prefix is "R." OK.

Under Setup > Real-time Settings choose Choice 1: Use Uncorrected GPS. OK.

Under Data, make sure the file type is Rover, then tap Create. If asked to Confirm Antenna Height put in 3 ft. OK. The receiver will begin taking data. Write down the file name, which will be of the form RMMDDHHA, where R is the file prefix, MM is the month, DD is the day, HH is the hour, and "A" indicates that this is the first file taken this hour ("B" would be the second, "C" the third and so on).

Begin your journey. The receiver will beep softly every time it takes a vertex. (The volume will depend on the windows setting.)

When you have completed your route tap Close and confirm that you want to close the file. Tap the "X" in the upper right of the screen and confirm that you want to exit TerraSync. Turn off the Juno ST.

ArcPad: You are going to collect the second of two folders of information. This will be of a sequence of vertices which will represent a linear feature such as a stream or road. Again its name will be of the form "QuickProject YYYY-MM-DD HHXX."

First be certain that GPS is active. Next you may want to specify that a vertex is taken only when the antenna is moved a certain distance. You'll find this under Capture under GPS Preferences, found under the GPS DdM (8A). If you are walking with the unit maybe you want to specify a distance of 10 feet. If on bicycle perhaps 30 feet. In an automobile you could use 200 feet. You will be "editing" a blank data set, so you will need the Edit toolbar. Bring that up with the Tools DdM (10A). The Edit toolbar appears, with every icon grayed out—because there is nothing to edit yet. But you will notice that the Command toolbar emerges at the bottom of the window.

To initiate the QuickProject you go to the Open Map DdM (2A) and tap New. Tap QuickProject. Read and OK the screen. If asked whether to save changes to Untitled say "No." If necessary uncheck "Keep layers from current map." OK. Write down the Project folder name, so you will be able to distinguish it from others in the datalogger. Notice that you have been provided three shape files: Points.shp, Lines.shp, and Polygons.shp. OK.

You are returned to the main window. Notice that parts of the Edit toolbar have come alive.

Tap the pencil icon (the Start/Stop Editing Pic (11B)) and you will notice that all three feature types are available for editing. You can only edit one at a time so turn off the red boxes around Points and Polygons.

Tap the Feature types DdM (4C) and select Polyline. Tap the icon Capture Multiple Points from GPS Pic (7C) and you will be presented with a screen asking for a Name for the line feature. Call it Line#1. Begin moving the Antenna.

You will notice that you don't see much on the map. That's because the vertices of the line are very close together and, as you can see from the scale bar, a lot of real estate is being shown. So use the Zoom In Pic tool (2B) and repeatedly drag a box around Line#1. Soon you will begin to see the line develop. If you zoom in too far, use the Previous Extent Pic (5B) or the Zoom Out tool (found on the Extents DdM) to back out some. You can also Zoom to Full Extent with 4B DdM. Observe the Scale Bar.

The line will develop as you move. The red circle with + in it is the current position that the GPS is calculating. ArcPad will attempt to keep it on the screen. As you move in directions that would take you off the map the current position will snap back to the center. You could prevent the window from always keeping the current position on the map by going to GPS Preferences (you remember where that is by this time) and clicking off Automatically Pan View (but don't). If you want to see the whole track click Zoom to Extents. Take a minimum of 200 points. When you have completed your route Stop recording data by starting a new QuickProject or exiting ArcPad with Open Map DdM (2A). If things didn't work out try again, using the summary of hints below.

Summary of Hints for Collecting Vertices Along a Line:

1. Turn on MM6 > Start > Today
2. Check date, fix if necessary
3. Start ArcPad
4. Cancel opening dialogue if it appears
5. Activate GPS Pic (7A)
6. Drag Position box to screen center
7. Set Datum to D_WGS_1984 (8A)
8. Set Display Units to Statute-U.S. Units (9A)
9. Disable Point Averaging (8A)
10. Set Distance Interval (e.g., 10 feet if walking, 30 feet if biking, 200 feet if driving) (8A)
11. Set Maximum PDOP to 8 and use 3D Mode Only (8A)
12. Set Antenna Height to 1 meter or 3 feet (8A)
13. Set all visual Alerts off, aural Alerts on (8A)
14. Turn off Status bar (10A)
15. Make sure positions are being calculated
16. Change Map Projection to DD (Tap Coordinates to change)
17. Display Edit toolbar (10A)
18. Start QuickProject (no current layers)(2A > New)

19. Enable Lines only with Pencil Pic (11B)
20. Enable Polyline with Feature DdM (4C)
21. Capture vertices from GPS Pic (7C)
22. Start Moving (DON'T TOUCH ANYTHING on the screen)
23. When done with the route: Proceed to Attribute Capture (If you don't do this you will lose everything!!!) (3D)
24. Give the Line a name
25. Start a new QuickProject **OR**
26. Exit ArcPad with Open Map DdM (2A)

See the various manuals for information on how to collect points along a path with your receiver and software. Record the information in your FastFactsFile.

{__3} *Stop recording data.*

{__4} *Complete the Data Collection Parameter Form.*

{__5} *Shut off the receiver.*

PROJECT 2C—UPLOADING DATA INTO A PC

Back Inside

{__1} *If you have not already done so, finish filling out the data collection parameter sheet(s) for the PROJECTS 2A and 2B data collection sessions.*

{__2} *If your receiver provides this information, compare the memory used and the number of fixes collected. How many bytes did the average fix require?* _____ .
If this is much larger than 25 then it may be that the unit had to repeatedly change constellations, and hence store more data, to maintain the set of four satellites with the best PDOP. While any extended data collection session will involve constellation changes—after all, each satellite is only visible from a given point of the Earth for a few hours a day—you will get the best data if you are careful not to force constellation changes by obstructing the signals.

{__3} *With the Receiver/Datalogger on and at a "home" screen with Start available, click Start > Settings > Connections. If you see "USB to PC," make sure Enable advanced network functionality is* **unchecked.**

Transferring Data from the Receiver to the PC

At this point you have recorded spatial data, but it is essentially locked away in the datalogger. Copying it from the datalogger into a personal computer (PC) is the next undertaking. This process is sometimes called **uploading data**.

The first step in transferring data is to make a physical connection between the receiver/datalogger and the PC. Your instructor will have paved the way for this. Probably all you will have to do is connect the datalogger to a USB cable, the other end of which is connected to the computer. Data transfer will require a program running on the PC. This may well be the Microsoft program called ActiveSync if you are using the Windows XP operating system (or its predecessors) and Windows Mobile (or its predecessors). If you are using the Windows Vista operating system this intermediate software will be called the Windows Mobile Device Center. There are other synchronizing programs, such as FinchSync, BirdieSync, and IntelliSync. Handheld receivers speak a different "natural" language than PCs, so it is not generally sufficient to simply copy data files from one to the other. The synchronizing software takes care of this data conversion. Setting up a synchronizing program can be tricky, with myriad pitfalls, so I'll assume it has been done for you or that you will find out how to do it for your combination of handheld and computer.

Probably your PC has been configured to accept data in the proper form. The local convention may be different, however. If things don't seem to be working, check with your instructor.

{__4} *Be sure the handheld is turned on. If you are using Windows Mobile make sure the handheld is on a "home" screen, showing Start in the upper left corner. Connect the handheld to the PC (probably with a USB cable*). When you complete the connection ActiveSync should start automatically. You will see it on the Windows XP or Vista task bar. If the connection does not work you can try a number of things. For example, do a hard reset on the handheld, or a reboot on the PC, or both.*

Once they are properly connected (meaning USB isn't complaining and ActiveSync is running) ActiveSync will suggest that you "synchronize" the PC with the handheld. You don't want to do that. So "X out" or cancel any windows that suggest that you want to actually synchronize the two devices. All you are interested in is transferring information between the devices. Use one of the three following procedures to transfer the data to the PC.

TerraSync and Pathfinder Office: These are both Trimble Navigation programs which communicate with each other, so ActiveSync runs in the background and, although it is used, it won't be seen during the process.

Execute the Pathfinder Office Software

{__5} **Start the Pathfinder Office Software:** *I'm assuming that you have a passing familiarity with Windows XP or Vista. If not, a crash course or some tutoring is in order. Your instructor will have given you some directions on how to log onto the computer you will use and to invoke Pathfinder Office. Follow those directions now.*

---

* USB stands for Universal Serial Bus.

{__6} *If a Select Project window came up when Pathfinder Office started, dismiss it (by clicking the "X" in the upper right of that window). The results should look something like Figure 2.10, perhaps missing a toolbar or two; we'll fix that. If the Pathfinder Office window does not occupy the full screen, double click on its title bar.*

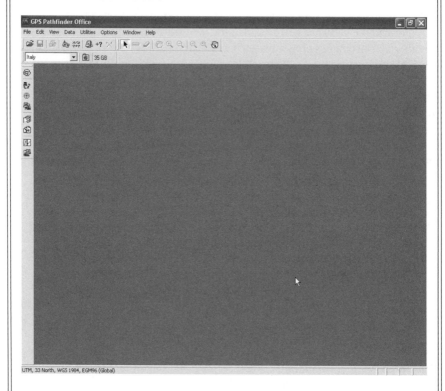

FIGURE 2.10   Main window of Pathfinder Office.

Project is one of those annoying English words that subsume two distinctly different pronunciations and meanings under a single spelling.* As a verb, it can mean, for example, shine light through film onto a screen, or *project* a piece of the Earth's surface to make a map—that is make an image or geographic *projection*. But *project*, as a noun, can mean a focus of activity, time, data, money, and so on so as to accomplish some goal, as in to "launch a project." In the case of Pathfinder Office, it means the latter, although, since the subject in general is geographic data, no one could complain had you guessed otherwise. In any event, the function of the Project is to serve as a closet for the files you will work with.

---

* Half a homonym, perhaps.

A Project in Pathfinder Office has several components. First is the Project Name. This name points to a primary folder (preferably with the same name but that is not required). The primary folder contains three sub-folders, named backup, base, and export.

{__7} ***Make a new Project with its associated folders:*** *In the File menu click Projects. (From now on I may indicate a sequence of actions like this by File > Projects.) Click "New" in the Select Project window. A "Project Folders" window will appear. You will make a project with the name DATA_yis, where "***yis***" indicates* **Y***our* **I***nitial***S** *(in capital letters, up to three). For example: DATA_MK. The project is to be located in a folder named DATA_yis, which will be under the folder ___:\MyGPS_yis, where "___" is "C," "D," "E" or whatever hard disk drive your instructor tells you the GPS2GIS folder is located on. Type*

DATA_yis

in the Project Name field. Type

___:\MyGPS_yis\DATA_yis*

in the Project Folder field. This procedure assures that the Project Name and the primary project folder window will be identically designated—which is good practice, but not required. See Figure 2.11.

If you click OK the window will disappear and the Select Project window will again become active.

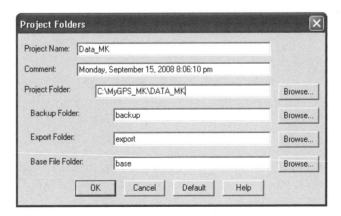

FIGURE 2.11   Project Folders window.

* Replace the "___" with whatever hard drive disk identifier your instructor tells you is correct.

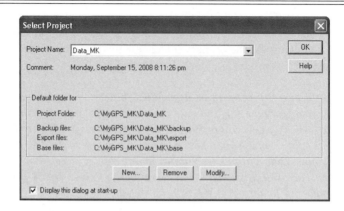

FIGURE 2.12   Select Project window.

{__8} *Examine the Select Project window. It should look about like Figure 2.12. If it does, OK the window. You have now set up Pathfinder Office so it will install your files in the proper places. You have established a main folder for your data, and three sub-folders under it for functions that will be explained later.*

{__9} *Select Options > Toolbars to make sure that checks appear next to all tool bar selections: Standard, Project, Mouse, and Utility. Also make sure a check appears next to the Status Bar menu item.*

{__10} *Check that the Project name appears on the Project toolbar. Also on this toolbar will be the amount of disk space remaining on the associated disk drive. You should also see a folder icon. Click on it. You will see the contents of the folder for the project. It will contain three folders: backup, base, and export. Dismiss the project folder.*

{__11} *In Pathfinder Office Click Utilities > Data Transfer. Use the drop-down menu to make the Device field say something like GIS Datalogger on Windows Mobile. To the right you should see the statement Connected to GIS Datalogger on Windows Mobile. If it doesn't, click the Add button. If it still doesn't try the connection again or ask your instructor. (See Figure 2.13.)*

{__12} *Press the "Add" button and pick Data File. An "Open" window will appear. After a bit of waiting, within it you will see the list of data files that reside within the datalogger. Move your mouse cursor over the buttons on the upper right of the window. Press the one whose tool tip says "Details." The files in the datalogger will appear in a list which you can alphabetize by clicking on "Name." Depending on the datalogger you will get different amounts of information. (See Figure 2.14.)*

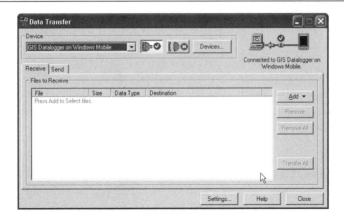

FIGURE 2.13   Data Transfer window.

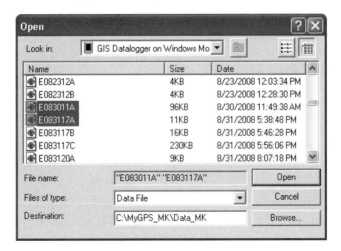

FIGURE 2.14   Designating which files to open.

   Select (with CTRL-Click) the two files that you created in Projects 2A and 2B (plus any other files you might have created during those sessions).

   (If you need to start over in selecting files, click in a blank area of the window.) Examine the Destination field to make certain the files are going where you want them to. Click Open. The "Open" window will disappear and the files you selected will appear in the Data Transfer Window. Click Transfer All.

   {__13} The transfer may all go by pretty fast for small files. (If the process should stop because of a communication error, just press the Transfer All button again. The transfer should proceed from where it halted.)

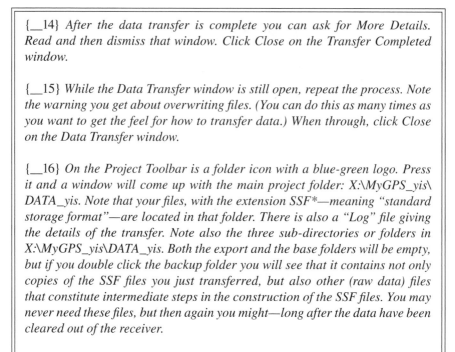

{__14} *After the data transfer is complete you can ask for More Details. Read and then dismiss that window. Click Close on the Transfer Completed window.*

{__15} *While the Data Transfer window is still open, repeat the process. Note the warning you get about overwriting files. (You can do this as many times as you want to get the feel for how to transfer data.) When through, click Close on the Data Transfer window.*

{__16} *On the Project Toolbar is a folder icon with a blue-green logo. Press it and a window will come up with the main project folder: X:\MyGPS_yis\ DATA_yis. Note that your files, with the extension SSF*—meaning "standard storage format"—are located in that folder. There is also a "Log" file giving the details of the transfer. Note also the three sub-directories or folders in X:\MyGPS_yis\DATA_yis. Both the export and the base folders will be empty, but if you double click the backup folder you will see that it contains not only copies of the SSF files you just transferred, but also other (raw data) files that constitute intermediate steps in the construction of the SSF files. You may never need these files, but then again you might—long after the data have been cleared out of the receiver.*

{__17} *Before you disconnect the handheld from the PC skip to the section below entitled **The Almanac**.*

* Depending on how Windows is set up, the extension "SSF" may or may not show in the folder.

ArcPad and ArcGIS: You will transfer QuickProjects from ArcPad to a folder named ___:\MyGPS_yis\Data_yis where you can look at the data with ArcGIS Desktop.

Assuming that Microsoft ActiveSync has come alive, open the ActiveSync window. It should indicate "Connected."

What you do want to do on the PC is to Explore the Device. Explore should show up as an option on a toolbar. If it doesn't, find it under Tools on the ActiveSync window. Clicking this will open a Mobile Device window that looks like you have opened a folder on the PC. Folders and Files on the handheld are shown. You should see a window which will contain your QuickProjects. If there are several QuickProjects, how do you know which is yours? Because you wrote down the date and time you took the data sets.

Now open the folder, on the PC, named MyGPS_yis. Make a subfolder named Data_yis. Tile the screen so you can see, side by side, both the window showing the files on the handheld and the window showing the empty folder Data_yis on the PC.

On the handheld window (Mobile Device window) pick out the two QuickProject folders that contain the data you just collected. Drag one of them from the handheld to your MyGPS_yis\Data_yis folder. When asked if you want to copy or move a file select "copy." (It is the best practice to copy a file from source "A" to destination "B," so that the file is duplicated. Once you are sure the copy has gone according to plan you can delete the file from source "A" if you wish. Using "Move" copies the file and then deletes it automatically from the source. If something goes wrong during the copy (a power failure, for example) you haven't lost anything by copy and a "manual" delete. If things go wrong in the midst of a move you might lose data.)

You will note something different during the copy from the handheld, than what you are used to in simply copying from one PC folder to another. You will get messages about data being converted. Again, the handheld and the PC speak different languages. Don't be concerned.

Disconnect the handheld. You now have the two QuickProjects on the PC as file folders. Those folders are within MyGPS_yis\Data_yis.

Read over the procedures above for the other receivers and PC software to get an idea of the objectives of this Project. Then see the various manuals that relate to your particular situation. Record the information in your MyGPS_yis file.

The Almanac

A NAVSTAR almanac provides a description of approximately where the satellites are at any given moment in time. Collected by and retained in the GPS receiver, the almanac tells the receiver which satellites it will be able to see during a data collection session. That is, the almanac helps in the initial phases of outdoor data collection. So the receiver automatically collects an almanac every time it is turned on, if it can. To collect an almanac, the receiver needs to be picking up the signal from at least one satellite for about 15 minutes. Almanacs are available from the satellites, all day, every day, and usually have validity for up to 3 months. Of course, if a new satellite is put up, or the existing ones are rearranged in their orbits, the almanac becomes less useful immediately.

The almanac may be copied from the receiver to the PC using Pathfinder Office. Having an almanac on the PC is not necessary in order to use the data you have collected, but later you will see that it is useful for planning a data collection session. For example, when used with a Pathfinder Office module called QuickPlan, it can tell you how many satellites are available at any time during the day, and what DOP values you can expect. Further, copying an almanac to the PC is illustrative of the communication process, so we do it for practice.

TerraSync and Pathfinder Office

{__18} *With the GeoExplorer still connected, and Pathfinder Office running, use the knowledge you gained from transferring files and set up Pathfinder Office to transfer the almanac to your PC. You will use the data type "Almanac." The path and Output File will appear by default probably as*

C:\Program Files\Common Files\Trimble\Almanacs\Almanac.ssf.*

Change the Almanac name to Almanac_yis and, after browsing, store it in ___:\MyGPS_yis\DATA_yis. (You can store the almanac on the PC by any name and in any place. However, unless told otherwise, the Trimble program QuickPlan will look for the latest almanac with the name "Almanac" in the default location.) When you press Transfer All, the Pathfinder Office will correctly claim that the almanac is being transferred.

{__19} *When the transfer is done, close the Data Transfer window and exit Pathfinder Office. Turn off the datalogger and disconnect it from the USB cable.*

You have completed uploading your data files and the current almanac. (You were asked to put the almanac in your own folder since others might want to use the "official" almanac usually located in

C:\Program Files\Common Files\Trimble\Almanacs\Almanac.ssf,

which might have been collected at a later time.) In the next chapter you will look at several sample data files with Pathfinder office, and then at your own files.

---

* "Program Files" and "Common Files" are both folder names. In what this author considers a significant design error, Microsoft Windows allows blanks to be present within folder and file names.

ArcPad and ArcGIS: The ArcPad software does not collect an almanac that can be downloaded to a PC.

See the various manuals for your software. If you can save the almanac do so and record the method in your FastFactsFile.

The Shape of the Earth—Considering Elevations

In terms of trying to find a mathematical or textual description of its shape, Earth is a mess. Even not considering its obvious bumpiness—clearly evident to you if you only look out a window—the overall shape defies any attempt at neat description.

The wonderful idea that Earth was the simplest of all three-dimensional figures—the sphere—bit the dust in 1687 when Newton proposed that an ellipsoid was a better approximation. Thus the Earth was considered to more resemble a ball, compressed slightly at the poles, with greater girth at the equator. The diameter through the poles is some 43 kilometers less than a diameter across the plane of the equator.

In the latter part of the 20th century it was determined that the shape departed from ellipsoidal as well. A slight depression at the south pole complements a little protuberance at the north pole, and just south of the equator we find a bulge. So how can a cohesive description be made? By careful mathematical work, based on gravity and water.

A definition: The **geoid** is a surface, like an egg shell. It is equipotential (that is, everywhere on the surface the strength of gravity is the same) and (almost) coincides with mean sea level. Imagine that all seas are calm, and mean sea level extends through the land forms through a network of canals. (In reality, mean sea level, itself, can vary by a meter or two, depending on where it is measured.) The geoid surface, by definition, is perpendicular to the force of gravity, no matter at what point it is considered. Because the value of gravity varies over the surface of the Earth the surface of the geoid is smooth and continuous, but not regular as a sphere or ellipsoid would be. It has depressions and bulges.

Satellites are kept in their orbits by gravity. Gravity can be considered a force between the satellite and the center of mass of the Earth. The GPS satellites are affected by neither the shape of the Earth nor variations in its density. They orbit around its center of mass and are a long way away. This provides the opportunity to define a simple mathematical surface that approximates the surface of the Earth. The **reference ellipsoid** is this surface—created by rotating an ellipse around the axis connecting the poles. The center of the ellipse, and hence the center of the ellipsoid, is the center of mass of the Earth. The surface of the ellipse is meant to approximate the geoid. Some places the reference ellipsoid surface is below the geoid and some places it is above.

Compare the Two Altitude Referencing Systems

For all but the last few years, most people measured altitude from the average level of the oceans. The two primary methods of taking measurements were vertical length measurement from a beach (inconvenient if no ocean were nearby) and measurement of air pressure. Of course, air pressure is variable from hour to hour so there are complications using this method as well.

As previously mentioned, a new definition of altitude has been developed, using not sea level as the zero but the gravitational surface called the reference ellipsoid. As I just said, the reference ellipsoid approximates mean sea level, but is slightly different from it almost everywhere. Since the satellites are slaves to gravity, the GPS receiver "thinks" of altitude and elevation in terms of **height above the reference ellipsoid (HAE)**.

Of course almost all data related to altitude, garnered over several hundred years, is expressed in altitude above mean sea level (MSL). So formulas and tables have been developed that indicate the difference between MSL and HAE. These are incorporated into the GPS receivers so they are able to display altitude in MSL. If you are using TerraSync, you will examine HAE and MSL in your area by doing the following:

TerraSync

{__20} *Turn the Juno GPS unit on. Recall that we made a waypoint called SinglePoint in the waypoint file named GPSWPF_yis (where yis represents your initials). That is a distinct, unique point in latitude, longitude, and altitude on the Earth's surface.* **The receiver kept in its memory the coordinates of that point you took in the field. The coordinates of this point can be displayed in various ways in the receiver.** *We can refer to it with a triple of numbers. The question will arise: which set of three numbers should we use in a given situation? Based on different datums, projections, coordinate systems, and units there are hundreds of triples of numbers that represent that same point. So our plan now is to look at different sets of coordinates and to understand that, if we combine data sets which use difference reference bases, we will get wrong answers. Let's look at altitude first. First you will determine the approximate difference between Mean Sea Level and the Height Above Ellipsoid in your area. (The manufacturer of the receiver coded the information in your receiver for all locations on the Earth's surface.) The idea is that you will first display the altitude in MSL, then display it in HAE.*

{__21} *Go to the Setup section. Tap Coordinate System. Make the System UTM. The unit will tell you what Zone you are in: _____. Make the Datum WGS 1984. Set the Altitude Reference to MSL and the Altitude Units to Feet, but set the Coordinate Units to those native to UTM: meters. OK.*

{__22} *Go to the Navigation section. Tap Waypoints. You should see the Waypoint File named GPSWPF_yis. Highlight that by tapping on the name. Tap Open. Now you see the waypoint you previously created named SinglePoint.*

{__23} *Go to the Map section. Tap the pointer arrow (Select) in the upper left. Now tap the crossed flags that represent SinglePoint. Notice that you get a "map tip" that tells you the name, the coordinates, and the altitude. Write down the altitude in the MSL blank below.*

SinglePoint Altitude (MSL) _____

SinglePoint Altitude (HAE) _____

Difference _____

{__24} *Go back to Setup > Coordinate System. Change the Altitude Reference to Height Above Ellipsoid (HAE). OK. Return to the Map section and again tap the waypoint. Write down the HAE SinglePoint Altitude in the blank above and subtract the lower altitude from the higher one. This is the difference from the Mean Sea Level altitude and the Height Above Ellipsoid altitude in your location. Which is higher? _____*

ArcPad: The ArcPad software does not display HAE altitudes.

If your GPS mobile or PC software can be set to display both MSL and HAE record how to do it in your FastFactsFile and fill out the table above.

The Datum Makes a Difference

It is absolutely vital, when integrating GPS data with GIS data, that your data sets match with respect to geodetic datum,* coordinate system, units, and projection. You will soon prove to yourself how important this is by filling out the Latitude and Longitude Computation Tables at the end of this chapter and noting the differences between different systems. You will need to use the following four points of information to determine the differences in position designation from one datum to another, and from one coordinate system to another:

- One degree of latitude corresponds to approximately 111 kilometers (km) or 111,000 meters.† (Just FYI, one minute of latitude corresponds to that number divided by 60, or approximately 1845 meters. And one second of latitude is about 31 meters.)
- The length of a degree of longitude, measured along a parallel, depends upon the **latitude** of that parallel. The length varies from approximately 111,000 meters at the equator to zero meters at the poles. So some computation is needed: at the latitude at which you took data (what is it? _____), one degree of longitude corresponds to approximately 111,000 meters **multiplied** by the **cosine** of that latitude. For example, if your latitude were 30° the value of the cosine would be approximately 0.866. Therefore a degree of longitude would correspond to 96,126 meters (that's 111,000 times 0.866 = 96,126). Now do the calculations for your latitude: At the position of your fix, a degree of longitude corresponds to _____ meters.
- In the UTM coordinate system, in a given zone, a greater number of meters indicates a more easterly position in longitude, or a more northerly position in latitude. That is, x and y increase "to the right" and "up," respectively, in accordance with standard Cartesian convention.

---

* *Datum* is the master geodetic referencing system used for a particular project or map.
† More precise numbers can be found by searching the Internet for *length* and *degree of latitude*.

Fill Out the Latitude and Longitude Computation Tables

TerraSync

{__25} *Turn the GPS unit on. Recall that the receiver kept in its memory the coordinates of the waypoint SinglePoint in the waypoint file GPSWPF_yis. Under Setup set the Coordinate System to Latitude and Longitude. Set the Datum to WGS 1984. Under Units set the Lat/Long Format to DD.ddd°. Using the same method as you did determining the difference between MSL and HAE, write the coordinates in the appropriate blanks of Latitude and Longitude Computation Tables below. Now switch the datum to NAD 1927 (Conus) and write the coordinates.*

{__26} *To complete the table you need to obtain four more numbers. These are the northing and easting numbers using the UTM system with both the WGS 1984 datum and the NAD 1927 datum. So repeat the process above, but with the coordinate system set to UTM.*

When you are finished writing down the coordinates and doing the calculations on the forms, come back and answer the question below:

{__27} *Does the difference in meters from WGS-84 to NAD-27 using the UTM coordinate system correspond to the difference in meters you calculated based on latitude and longitude degrees and minutes? _____ What conclusions can you draw from your observations and calculations? _____*

_____

_____

_____

ArcPad: ArcPad does not support the display of coordinates with different datums. So we will do this exercise when the data sets you collected are put into ArcGIS, which will let us look at the data with different parameters.

See the various manuals for your GPS unit. If you can display all datum/lat-lon/ UTM combinations fill out the tables below. Otherwise wait until we can put the data sets into ArcGIS.

LATITUDE (NORTHING)*

COMPUTATION BASED ON A SINGLE WAYPOINT OR FIX

~~~~~~

Latitude (Decimal Degrees)

NAD-27 _____ (N-Am. 1927)

WGS-84 _____

Difference _____

In your area, the representation in Decimal Degrees of the point in WGS-84 (alias NAD-83) is
_____ degrees (north or south) _____ of the representation of that point in NAD-27.

After computation as indicated in the text, this corresponds to _____ meters.

~~~~~~

Northing—UTM (meters)

NAD-27_____

WGS-84_____

Difference _____

In your area (which is UTM Zone: Number ___), the representation of the point in UTM
coordinates based on WGS-84 is _____ meters (north or south) _____ of the representation of that point in UTM NAD-27.

~~~~~~

---

* Microsoft Word Doc versions of all forms in this book may be found on the accompanying CD-ROM
  in the folder named FORMS.

LONGITUDE (EASTING)

COMPUTATION BASED ON A SINGLE WAYPOINT OR FIX

~~~~~~

Longitude (Decimal Degrees)

NAD-27_____

WGS-84_____

Difference _____

In your area, the representation in "Deg & Min" of the point in WGS-84 (NAD-83) is
_____ minutes (east or west) _____ of the representation of that point in NAD-27.

After computation as indicated in the text, this corresponds to _____ meters.

~~~~~~

Easting—UTM (meters)

NAD-27_____

WGS-84_____

Difference _____

In your area (which is UTM Zone: Number ___), the representation of the point in UTM
coordinates based on WGS-84 is _____ meters (east or west) _____ of the representation of that point in UTM NAD-27.

~~~~~~

Data Collection Parameter Form—GPS2GIS

Memory Remaining in the Datalogger _____K

Hours of Battery Use Prior to this Data Collection _____:_____

Location _____

_____

_____

Date _____ Day _____

Filename _____

Time: Start _____

Using SVs (initially): _____ _____ _____ _____

PDOP _____ (HDOP _____ VDOP _____ TDOP _____)
                    (The GeoExplorer displays only PDOP)

Position Logging Interval: 1 fix per ____ (number) _____ (units)

Total Number of Fixes _____

Time: Stop _____ Duration _____

Using SVs (at end): _____ _____ _____ _____

PDOP _____ (HDOP _____ VDOP _____ TDOP _____)

File Size (in K) _____. File Size (in Bytes: K*1024) _____Bytes

Bytes per Fix (Number of Bytes divided by number of Fixes) _____

Memory Remaining in the Datalogger _____K

Hours of Battery Use _____:_____

Notes: _____

_____

_____

3 Examining GPS Data

*In which we continue our discussion of
the theoretical framework of GPS position
finding, and you practice using PC software
to investigate files collected by GPS receivers.*

OVERVIEW

SOME QUESTIONS ANSWERED

As you read the last two chapters some questions may have occurred to you. And the answers to these questions may generate other questions. Here are some that come up frequently.

Question #1: The captain of the ship of Figure 2.1 had a map showing the locations of the soundhouses. But how does the GPS receiver know where the satellites are?

A map is a two-dimensional scale model of the surface of the Earth. But models can take many forms, including mathematical. Due to the nature of nature, as elucidated by Isaac Newton and Johannes Kepler, the position of a satellite at any time may be predicted with a high degree of accuracy by a few mathematical equations. A satellite orbiting the Earth may be modeled by formulas contained in the memory of the microcomputer in the receiver. When the formulas are applied to bodies at the high altitudes of the GPS satellites, where they are free from atmospheric drag, the formulas are relatively simple and can predict the position of the satellite quite accurately.

Almost all formulas have a general form, into which specific numbers are "loaded." For example, in an equation of the form

$$Ax + By = z$$

A and B are **parameters** that represent constant numbers that may be inserted in the equation. When A and B are replaced by actual numbers, then the equation is only true for certain values of x, y, and z. The receiver carries the general form of the formulas that give the position of each satellite. Before the range readings are taken by the receiver, the satellites will have broadcast the values of their particular parameters so the receiver can complete its equations. Then, by knowing the current time at a given moment (the moment at which the distance reading is taken), the receiver can know where the satellites are.

Actually, the satellite message coming to the receiver antenna is in many parts. Two of these might be called the almanac and the ephemeris data. **Almanac** information is broadcast to provide close, but not precise, satellite position information. The almanac for all satellites is broadcast from each satellite. Further, each satellite broadcasts **ephemeris** information (which applies to that satellite only) that provides up-to-the-minute corrections. The satellites are not completely predictable in their orbits because of such forces as gravitational pull from the Sun and Moon, the solar wind, and various other small factors. Therefore the satellites are carefully monitored by ground stations and told their positions; each satellite then rebroadcasts this information to GPS receivers.

Question #2: The captain needed to know exactly what time it was in order to determine his distance from the soundhouse. How is the clock in the receiver kept accurately on GPS time?*

The short answer is that the receiver clock is reset to GPS time by the satellites each time a position is found. Such resetting is necessary because, while the receiver clock is very consistent over short periods of time, it tends to drift over longer periods. (Each of the four atomic clocks in each satellite cost about $50,000 each; the single clock in the receiver obviously costs a whole lot less, so you can't expect the same sort of accuracy. If you don't use the receiver for a week or two, you may notice a difference of several seconds between the time the receiver displays and true time.) The clocks in the satellites keep time to about a tenth of a billionth of a second (a tenth of a nanosecond).

If you consider "time"† as the fourth dimension and accept that it takes one satellite to fix each dimension, then it is clear that four satellites, working in concert, can set the clock and provide a 3D spatial position.

As you may recall from our discussion of the theory of GPS and from the geometry of the diagrams you examined, you might presume that only three satellites are required for a 3D fix. But given that the receiver has only an approximate idea of what time it is, what must be calculated is a 4D fix. So four satellites are required. It is not correct to say that three satellites are used for the 3D fix and the fourth sets the receiver clock. Rather, all of the satellites operate in concert to find the true "position" of a receiver antenna that may move in space (relative to the Earth) and does move in time.

(GPS, as previously mentioned, has had a revolutionizing effect on the business of keeping extremely accurate time—to better than a billionth of a second. Although most of those who use GPS are concerned with finding positions, the system also supplies extremely accurate time signals to receivers whose positions are known with high precision. GPS has made it possible to synchronize clocks around the world. This has made it possible, among other things, to gain knowledge about the makeup of the Earth's center. Because seismologists throughout the world know the exact time, they

---

* GPS time is almost identical to UTC time. At one time they were identical, but the world's timekeepers, starting in 1972, have had to insert a leap second occasionally to keep solar noon and the rest of the hours where they ought to be. The Earth's rotational velocity is slowing because of tidal breaking (oceans sloshing around). Leap seconds are added every year or two. But the basis for GPS time does not change. At the turn of the century UTC time was exactly 13 seconds ahead of GPS time.

† Humorously described as nature's way of keeping everything from happening at once.

can track the shock waves from earthquakes as they pass through the Earth. Another use of the ability of GPS to synchronize clocks around the world is in routing Internet (World Wide Web) traffic. It would not have been unreasonable to have called GPS by the acronym GPTS: the Global Positioning and Timing System.)

Question #3: The soundhouse sent a signal every minute. How often does a satellite send a signal? What is the signal like?

Actually, each satellite sends a signal continuously, rather like a radio station that broadcasts 24 hours per day. The radio station signal could be considered to consist of two parts: a carrier, which is on all the time, and "modulation" of that carrier, which is the voice or music which you hear when you listen to the station. (You probably have detected the presence of the carrier when the people at the station neglect to say or play anything.* The carrier produces silence (or a low hiss), whereas if your radio is tuned to a frequency on which no nearby station is broadcasting you will hear static.)

Each satellite actually broadcasts on at least two frequencies. Only one of these is for civilian use. (The military units receive both.) The civilian carrier frequency is 1575.42 megahertz (1,575.42 million cycles per second). In contrast, FM radio signals are on the order of about 100 megahertz. So the GPS radio waves cycle about 15 times as often, and are, therefore, one-fifteenth as long: about 20 centimeters from wavetop to wavetop. As this goes to press there is serious discussion about adding one or two new civilian signals. Having two signals at different frequencies available allows a receiver to compute a more accurate position than does a single signal.

The modulation of the GPS wave is pretty dull, even when compared to "golden oldies" radio stations. The satellites broadcast only "bits" of information: zeros and ones. For most civilian use, this transmission, and the ability to make meaning out of it, is called the "C/A code"—standing for Coarse/Acquisition code. The word "Coarse" is in contrast to another code used by the satellites: the "P" or "Precise" code. The term "Acquisition" refers to the capability that allows both civilian and military receivers to acquire the approximate position of the receiver antenna. The C/A code is a sequence of 1023 bits which is repeated every one-thousandth of a second.

A copy of the C/A code might look like this:

$$1\ 0\ 0\ 0\ 1\ 1\ 0\ 1\ 0\ 0\ 1\ 0\ 1\ 1\ 1\ 1\ 0\ 1\ 1\ 0\ 0\ 0\ 1\ .\ .\ .$$

and on and on for a total of 1023 bits. Then the sequence starts again. The sequence above probably looks random to you—as though you began flipping a coin, recording a "1" each time it came up heads and a "0" for tails. It is, in fact, called a **pseudo-random noise** code—the term "noise" coming from the idea that an aural version of it would greatly resemble static one might hear on a radio. The acronym is **PRN**.

---

* Dead air.

Question #4: How does the receiver use the 0's and 1's to determine the range from the satellite to the receiver?

The PRN code is anything but random. A given satellite uses a computer program to generate a particular code. The GPS receiver essentially uses a copy of the same computer program to generate the identical code. Further, the satellite and the receiver begin the generation of the code at exactly the same moment in time.

The receiver can therefore determine its range from the satellite by comparing the two PRN sequences (the one it receives and the one it generates). The receiver first determines how much the satellite signal is delayed in time, and then, since it knows the speed of radio waves, it can calculate how far apart the two antennas are in space.

As an example (using letters rather than bits so we can have a more obvious sequence, and cooking the numbers to avoid explaining some unimportant complications), suppose the satellite and the receiver each began, at 4:00 p.m., to generate one hundred letters per second:

$$\text{G J K E T Y U O W V W T D H K} \ldots$$

The receiver would then look at its own copy of this sequence and the one it received from the satellite. Obviously its own copy would start at 4:00, but the copy from the satellite would come along after that, because of the time it took the signal to cover the distance between the antennas. Below is a graphic illustration of what the two signals might look like to the computer in the receiver:

```
4:00 p.m. >>↓                    ↓<<<<<<7/100 of a second after 4:00 p.m.
Receiver:    G J K E T Y U O W V W T D H K . . .
Satellite:               G J K E T Y U O W V W T D H K
```

The receiver would attempt to match the signals. You can see that the signal from the receiver began to arrive seven letters later than 4:00; the receiver's microcomputer could therefore determine that it took 7/100 of a second for the signal from the satellite to reach the receiver antenna. Since the radio wave travels at about 300,000 kilometers per second, the time difference would imply that the satellite was 21,000 (that is, 7/100*300,000) kilometers from the antenna.

Question #5: The receiver must find ranges from at least four satellites to determine its position. How does the receiver "listen to" several satellites at once? Because all satellites broadcast on the same frequency, how does the receiver identify the satellites?

The first thing to know is that each satellite has its own distinctive PRN code. In fact, the satellite numbers you were logging in the first assignment were the PRN numbers—which is the principle way satellites are identified. A satellite may also have a number painted on its side, but it is the PRN number that counts. When an older satellite is retired, its replacement can take on its PRN number.

Most receivers have several electronic components, called "channels," that are tuned to receive the civilian GPS frequency. Although all channels are tuned to the same frequency, a single channel can track a GPS satellite by locking onto its PRN code. In more expensive receivers with several or many channels, each channel is assigned full time to tracking a single satellite. Older receivers "time share" a channel—flipping it between satellites, as you might flip between channels on a TV, trying to keep track of two programs at once.

Question #6: I've heard that the accuracy of GPS receivers was greatly increased when selective availability was turned off. What was selective availability? Why did it exist?

Selective Availability, or **SA**, was the error deliberately introduced by the GPS managers in the C/A code broadcast to diminish the accuracy of GPS receivers. Sometimes the satellites lied about their positions. Sometimes they lied about when they sent the code.

What was the extent of the error caused by SA? The government guaranteed that 95% of the time a fix would be within 100 meters of the true position. To understand why SA existed you have to realize that the NAVSTAR system started as a military project to provide navigation for units of the armed forces. In the broad sense, GPS was designed as a weapons support system. One doesn't want one's weapons to fall into enemy hands. So steps were taken to deny use of the system to all but authorized receivers. In fact, the very existence of the GPS system, whose first satellite was launched in 1978, remained secret for several years.

It was never planned that you could buy a $200 receiver for your fishing boat. The military feared such uses as a terrorist with a mortar knowing exactly where he was, and hence being able to more accurately target his fire. Or the computer in a missile being able to monitor its position and correct its path during its flight.

If sufficient warning were given, of course, the entire civilian side of NAVSTAR could be shut down to deny its use to hostile forces. (The consequences would be disastrous, but not so much so as a nuclear war.) But, even so, the military was still uncomfortable with allowing the best GPS accuracy in the hands of everybody. So why was SA turned off? As it turns out, very good accuracy may be obtained by using two GPS receivers in concert (we explain this in detail in Chapter 4) and for the very best accuracy you need two receivers, SA or no. So SA became more of a nuisance that offered no real protection. In fact, the Army Corps of Engineers began broadcasting corrections to positions obtained by civilian receivers. Under pressure from the civilian GPS users, and other countries—Japan and European—who began contemplating their own version of GPS, using the more general term Global Navigation Satellite System (GNSS),* the U.S. government abruptly clicked the SA switch off. This occurred on 2 May 2000, just after midnight Eastern Time. It was an important enough decision that the President made the announcement himself. The slim protection SA provided might be replaced and enhanced by jamming the GPS signals in selected geographical areas, if necessary. If you are interested in SA, and other

---

* At this writing, the Europeans are currently developing a particular GNSS called Galileo.

matters related to GPS policy, you can go to the Web site of the Interagency GPS Executive Board (http://www.igeb.gov) and find out considerably more than there is space for here. You can also view the President's declaration of 1 May 2000.

Question #7: How is it that a satellite, cutting Earth's meridians at 55° and moving at 8600 mph, generates a track that is almost due north-south in the vicinity of the equator, as seen in Figure 1.1?

While the satellite is moving very fast, it is also far out in space. Therefore the motion of the corresponding point on the Earth's surface along the satellite's track (picture where a line from the center of the Earth to the satellite would intersect the surface of the Earth) is considerably slower—about 2,100 mph. The satellite's track along the surface of the Earth moves at this speed toward the northeast on the upswing and southeast on the downswing, so the eastward part of its motion is in the same direction as the rotation of the Earth. Any given point on Earth's surface at the equator moves about 1050 miles per hour eastward due to Earth's rotation about its axis. The north or south component of the satellite's velocity is about 1,700 mph, while the east component is only about 1200 mph. So an observer at the equator would see only a slow drifting (about 150 miles per hour) of the satellite to the east over the period of an hour or two.

Question #8: If the orbital period is 12 hours, why does each satellite rise and set about 4 minutes earlier each day? Could the NAVSTAR system designers arrange to have the same satellites in view at the same time each day in a given location?

The short answer to the first question, if you know a bit of astronomy, is that the satellites orbit the Earth twice during a **sidereal** (pronounced *si-dear-e-ul*, meaning "star-based") day, rather than a **solar** (Sun-based) day. A longer explanation: Suppose you look up at the stars on midnight of the first of April and note their positions. To see the same picture on May first you have to look up at 10:00 p.m. In the 30 days the stars "moved" 120 minutes—4 minutes a day. Of course the stars didn't move; the Earth rotated. By midnight it would have turned not only the 10,800 degrees (that is, 360 per day times 30 days) from its daily rotation, but 30 degrees further (360 degrees multiplied by one-twelfth) from the April portion of its orbit around the Sun. The 10,800 degrees works out to zero basically (you are back where you started from) but the 30 degrees is significant.

Or think of it this way: While each individual satellite is in orbit about the Earth, making its circuit exactly twice a day, the set of satellites are independently in orbit about the Sun. They orbit the Sun as a package—sort of like a spherical birdcage made up of rings that are the orbits. The cage is centered on the Earth and contains it. The Earth is a body that rotates independently within this cage. The cage does not rotate at all on its own axis, but orbits the Sun.

To understand what "not rotating on its axis" means, realize that the Moon rotates on its own axis once during each trip around the Earth, so that it always shows the same face to Earth. If the Moon did not rotate, we would see different sides of it as it made its way around the Earth. In contrast to the Moon-Earth situation, consider the cage-Sun situation: the cage *does not* rotate on its axis and therefore does present different sides of itself to the Sun over the course of a year.

So at any given time (say noon, when the Sun is directly over a given meridian), a person on Earth will see (that is, "look through," toward the Sun) one side of the cage on the solstice in January. But from the same point on Earth that person would be looking through the opposite side of the cage in July. In effect, then, the cage will be seen from Earth to have made half a complete rotation around the Earth once each half year. To a person on the Earth, then, the cage apparently moves about 1/365th of a rotation per day. That amounts to about 4 minutes a day—calculated as 1440 minutes in a day divided by the number of days in a year.

In answer to the second part of question number 8, if the satellites' orbits were boosted another 50 kilometers or so further out they would appear in the sky at the same place at the same time each day. The further a satellite is from the Earth the longer its period—both because it moves more slowly and because it has further to go.

Question #9: In earlier text it was suggested that it was somewhat more important that there be a good view of the sky to the south for good reception. Why?

The statement about reception being better toward the south applies only to the middle and upper latitudes in the northern hemisphere. As you know, the satellites are in oblique orbits. Their tracks give the north and south poles a wide berth. In Figure 3.1 you are looking directly down on the North Pole at satellite tracks generated over a 6-hour period. The dashed circle is a parallel at 45°, so you can see that there is a dearth of satellites overhead if you go very far north of that. There is still good GPS coverage—all the way to the North Pole.

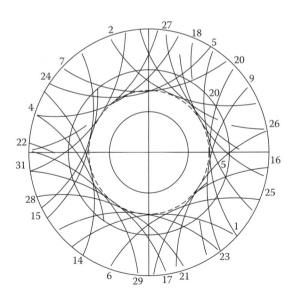

FIGURE 3.1 GPS satellite tracks seen from space look toward the North Pole.

Recall the story of the two "logically challenged" people who rented a boat and went fishing. They were highly successful—catching a lot of fish. Said "A" to "B," "Be sure to mark this spot so we can come back to it." As the day ended and they were approaching the dock, "A" asked "B": "Did you mark that fishing spot?" to which "B" replied: "Sure I did, just like you asked me—right here on the side of the boat." There was a pause as the absurdity of this penetrated "A's" brain. "You idiot! What if next time we don't get the same boat!"

But now you have a way of marking where you were. In fact, you have done so. In Chapter 2 you uploaded files from the GPS datalogger into a PC. By applying the programs we are about to use, you can determine where you are to within 5 to 15 meters. Our goal now is to look at those files, and some others, both graphically and statistically. As we begin the Step-by-Step section below (you need to pick the one that is right for your software) you will also acquire an understanding of the quality of your data.

CHAPTER 3A: STEP-BY-STEP

> If you are using Trimble Pathfinder Office, proceed with Project 3A. The text below serves as a tutorial for Pathfinder Office, as well as illustrating GPS technology. The files you will be using are in Trimble's Standard Storage Format (SSF).

> If you are using only ESRI ArcGIS software, skip to **Chapter** 3B Step-by-Step.

Project 3A—Volcano

We begin by looking at a file generated by a GPS receiver that circumnavigated Kilauea Caldera, the active volcano on the island of Hawaii. Before you can see or analyze the data, however, several parameters must be set up. (Right. Just as when you used the GPS receiver. It turns out that, when you do GPS for GIS, a considerable part of the activity is getting settings right. Sorry.)

{__1} *From the CD-ROM that came with the book, copy the folder GPS2GIS to a hard drive (such as C:) on your computer. From now on I'll make reference to this folder as ___:\GPS2GIS, leaving you to fill in the drive letter (or perhaps the drive letter and a path) to the folder.*

{__2} *Start the Pathfinder Office software as you did in Chapter 2. Close all open windows (except the main one, of course). Maximize the window.*

{__3} *In the File menu, pick Projects. Click on the field next to "Project Name." Find "Default" in the drop-down list (you may have to scroll the list up or down to see all the entries) and click on it. Okay your choice.*

{__4} *Under the "Options" menu, select "Coordinate System." You want to "Select By" "Coordinate System and Zone" (rather than "Site"). The "System" should be "Latitude/Longitude" and the "Datum" "WGS 1984." Altitude should be measured from mean sea level in meters (use the EGM 96 (Global) Geoid). "OK" your choices.*

{__5} *Under "Options ~ Units" pick Kilometers for the Distance measurement.*

{__6} *Under "Options ~ Time Zone" pick "Hawaii." Notice that Hawaiian time is 10 hours earlier than Greenwich (UTC) time.*

{__7} *Also under "Options," turn off all five "bars" (four "Toolbars" and the "Status Bar") by clicking each so that the check beside the toolbar name disappears. Now turn on the Standard toolbar. Run the mouse pointer over it, pausing over each button long enough to read the explanatory box that appears. Do the same now for the Mouse toolbar. Add and checkout the Project toolbar. Likewise the Utility toolbar. Finally, make the Status Bar appear at the bottom of the window. (Part of the trick to operating complex software is reading and understanding various messages that it provides— that is, keeping up with the status of the program. Another part is knowing the major capabilities—illustrated here by various buttons on the toolbars—of the software.) The completed window will look like Figure 2.10 in the last chapter.*

{__8} *Under Options, instruct the software **not** to save the window layout when the program is exited.*

{__9} *From the File menu, click on "Open." A window entitled "Open" will appear. (To achieve the same effect, you could also have clicked on the little yellow file folder icon at the left of the standard toolbar.)*

{__10} *Fill in the blanks in the window so that the folder is ___:\GPS2GIS\HAWAII* and the file is "Volcano.ssf." (The following step contains detailed instructions.)*

{__11} ***Navigate to the correct file:***
    (1) *click on the little down arrow in the "Look in:" box and, from the drop-down list with a mouse click, pick the choice that says something like "My Computer" or "This Computer." (Use the slider bar or arrows at the right of the window to see all the choices if necessary.);*
    (2) *within the large white area of the "Open" window select the icon associated with the disk drive "___:" with a single mouse click (it will become highlighted) and tap the "Enter" key (or double click on the icon, which comes to the same thing);*
    (3) ***pounce***† *on "Gps2gis" from the list that appears (you may have to use the slider bar at the right to get to it); alternatively use the down and up arrow keys to move the highlight;*

---

* Place in the blank whatever hard drive disk identifier your instructor tells you is correct.
† From now on, we'll call a single click on a choice, followed by the "Enter" key, a **"pounce."** You can also "pounce" with a double click but there are reasons not to do this if you aren't too familiar with Windows and/or the application software. For one, double click speeds are user selectable and a previous user may have selected a speed that makes your double click not work and leaves you wondering what happened.

(4) *pounce on "Hawaii;" (if you pounce on the wrong item you can backtrack by clicking the file folder with the "up arrow" next to the "Look in:" box);*

(5) *the "Files of Type" box should say "Data files (ssf, cor, phs, imp)" (If it doesn't, select this option from the drop-down list.);*

(6) *under "File Name:" select (with a single click) "Volcano.ssf" but don't open the file yet. Note, from the information at the bottom of the "Open" window, that the data collection start time was approximately 2:50 p.m., local time, on February 15, 1994. The file is made up of 110 positions (points, fixes) and occupies 12.9 kilobytes (KB) of storage. (If it doesn't say this, you did something wrong above. For example, if you picked the incorrect time zone, the date and/or time might show up incorrectly.) Now click the "Details" button (it's in a drop-down menu to the far right of the "Look in:" box) to see information about this and the other files in this folder. You can see file name, size, type, date the file was last modified, and attributes—either by moving the horizontal slider bar or resizing the column widths by dragging their separator bars.*

{__12} *Click "Open."*

{__13} *Display the map by clicking "Map" in the "View" menu. It may look something like Figure 3.2. Or it may not. Or it may be that nothing appears in the window at all. No matter! You have a lot of control over the appearance of the map, as you will see in the next few steps.*

{__14} *From choices that show up under "View," select "Layers" then "Features." The window that shows up (Feature Layers) contains a list of feature-type names.*

The data you are working with consists of simple position data (fixes consisting of a latitude, a longitude, an altitude). In Pathfinder Office such raw data are called "Not In Feature" data. (Positions may also be collected in such a way as to be associated with particular features, such as roads or types of vegetation, and these features may have names; I'll discuss the collection of such "feature attribute data" later in detail.)

{__15} *To change the way raw-position data appear on the screen map, single click on "Not In Feature" so it is highlighted. A box next to the name shows the symbology with which data are displayed. A checkmark, toggled with a mouse click in the box in the "Show" column, determines whether the layer will be visible or not. Make certain that "Not In Feature" has a checkmark beside it. The data you are working with consists of individual points that are not part of any other feature—hence the "Not In Feature" designation. (Make sure the "View" option in this window is set to "As above.")*

{__16} *In this same window, press "Line Style." A "Not In Feature" window appears. Here you can select a symbol color, a line thickness, and whether or not you want adjacent data points connected with line segments. Pick a medium line of red color with the data points* **not** *joined. (See Figure 3.3.) Okay your choice.*

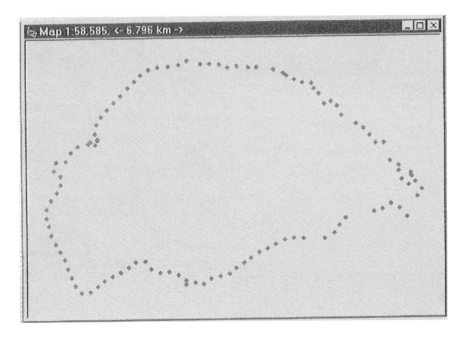

FIGURE 3.2 GPS track around Kilauea Caldera volcano.

{__17} *Verify that your choices of "Line Style" made it to the "Features Layers" window. Click OK there. Now the map should look like Figure 3.2 (except for the color). (Don't "Maximize" this window; if you do, the map scale and distance information bar at the top of the window will disappear!)*

Explore the Map

{__18} *What map scale appears on the map?* _____.

A map scale is a fraction; it is usually much less than one (1.000); its form is "1 divided by x, where "x" is usually a number much larger than one. If the number "x" is one (1.0000) the map is said to have full scale. Other ways of putting it would be that the map scale is 1.0/1.0, or 1:1, or 1 to 1—all of which imply full scale. Farmers would object if you began unrolling such maps. As the scale gets smaller (nearer zero) less detail is displayed and more real estate is shown on a given size page. Many people reverse the meaning of large and small scale; they look at the denominator of the fraction, rather than the fraction itself. (A 1:250 scale map is a larger scale map than a 1:500 scale map. One divided by 250 is 0.004; one over 500 is 0.002. The number 0.004 is larger than the number 0.002.) Remember it this way: larger scale maps have a larger degree of detail.

{__19} *Does the number of kilometers indicated at the top of the map refer to the distance from one side of the track to the other, or to the width of the window?* _____ (Hint: resize the east-west dimension of the window.)

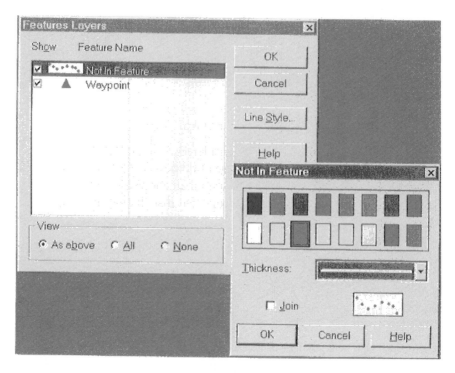

FIGURE 3.3   Defining feature-type symbology.

{__20} *Experiment with different types of "Distance" units from the "Units" window, created by selecting from "Options ~ Units." Note the change in the value of the east-west dimension of the window (shown in the map border at the top) reflecting that different units have been specified. Try inches.* Finish up by selecting "Meters" as the distance units for the data set display.*

{__21} *Since you have resized the window, you may want the image to fill it. Select View ~ Zoom ~ Extents and note the results. Now resize the window again so that part of the figure is cut off. This time find the magnifying glass icon with the equal sign (=) in it; press it. Again you get a "Zoom to Extents."†*

{__22} *In the "Units" window, set the "Area" to be "Hectares." (A hectare is 1/100 of a square kilometer. A hectare contains the same area as a square that is 100 meters on a side—but of course it might have any shape; there are about two and a half acres in one hectare.) Set the "North Reference" to be "Magnetic" (Automatic*

---

* The map scale shown is only approximate. Depending on the size of your monitor it may be off quite a bit. You can check by making the window 10 inches wide (use a ruler), multiplying the value in the denominator of the scale shown by 10, and seeing if this agrees with the distance given at the top of the window.

† As with many operations, there are multiple ways to achieve the same state. I won't point them all out. Once you become comfortable with the operations of Pathfinder Office you can look at the buttons and determine what shortcuts might be useful. Some of the menu options have buttons; some don't.

*Declination), so Pathfinder Office will automatically calculate the **declination** (i.e., the difference in degrees between true north and magnetic north).**

The map will show features oriented toward true north regardless of whether "True" or "Magnetic" North is selected. However when the Measure capability of Pathfinder Office is used, the direction of measuring lines will be shown in True or Magnetic degrees, depending on which is selected.

{__23} *Under "Options ~ Style of Display" choose the "Lat/Long Format" DD*MM'SS.ss" which will cause the software to display latitude and longitude coordinates in the form of degrees followed by minutes, followed by seconds including decimal fractions of seconds.† The other forms of display are DD*MM.mmm' and DD.ddd. Could you express 49°30'30.66" (49 degrees, 30 minutes, and 30.66 seconds) in these other two forms? There are 60 minutes in a degree; there are 60 seconds in a minute.‡*

(1)_____   (2)_____

{__24} *Now under "Options ~ Style of Display" change the "Lat/Long Format" to DD*MM.mmm'. Push the "Select" button (its icon is an arrow, pointing about north-north-west). Move the pointer around the map and observe the change in the "minutes" value shown on the Status bar. Recall that a minute of latitude is about 1845 meters. In general, the Status bar will display the coordinates of the map position indicated by the pointer.*

{__25} *Change the style of display to "DD.ddd."*

{__26} ***Determine some data point coordinates:*** *Slide the pointer down over the map. Note the status bar (at the bottom of the screen): the numbers there tell the latitude and longitude of the end of the pointer in decimal degrees. Position the pointer near the southwest portion (that's where the actual crater is) of the "oval" made by the GPS track. Write down a latitude and longitude: _____ If you click near a point on the GPS track, an "X" will appear, jumping to the nearest data fix. Try this. If you move the pointer off the map window the status bar will tell you the coordinate system in use. Try this.*

A number of Internet sites allow you to display maps at various scales. At the time of this writing (and bear in mind that World Wide Web sites are constantly in flux) there are sites that allow you to put in latitude-longitude coordinates and obtain a map of the surrounding territory. Some such sites are Google Earth, MapQuest, Microsoft Virtual Earth, and others.

---

* The Lat-Lon graticule is based on the location of the axis about which the Earth spins—and that axis defines the **true** north and south poles. The **magnetic** north pole is the place toward which compasses point. It is in very northern Canada at approximately 75° north latitude and 100° west longitude.

† The degree symbol is not particularly easy to produce on a computer, so Pathfinder Office lets you substitute an asterisk (*).

‡ To convert degrees, minutes, and seconds to decimal degrees, divide the seconds by 60, add the result to the minutes. Divide that sum by 60, then add this last quotient to the number of degrees.

{__27} *If you have access to the Internet, minimize Pathfinder Office and go to a mapping site and find where you can get a map and/or aerial view based on the latitude and longitude of the vicinity. Type in the latitude and longitude values you wrote down above (remembering that* **west** *longitudes are* **negative** *numbers). If the map-serving people have done their job correctly you should see a map of the road around the volcano (Crater Rim Drive). The central location indicated on the map should be approximately the location indicated on the Pathfinder Office portrayal of the GPS track. Note the similarity with the GPS track.**

{__28} *Back in Pathfinder Office with the Map window active, using the mouse, run the pointer along the icons on the mouse bar. Usually an explanatory box appears if you pause briefly over an icon. Find the one that allows you to measure distances on the map. (It looks like a ruler; you can also find its equivalent "Measure" in the "Data" menu.)*

{__29} ***Measure some distances:*** *Click "Measure." If you now click any location on the map near a vertical edge and then drag the cursor around, you can see that the Status Bar displays the distance and direction of the line in the units you have set for the display. Measurement takes place to the little "+" sign cursor. Swing the measuring line around so that it is vertical, toward the north. Note that magnetic north is about 10 degrees to the east of true north. That is, to go directly north you would have to follow a compass course of 350°.*

{__30} *If you now click the left mouse button again you will anchor the moving end of the line and will see a display of the length and direction of the line. You can continue the measuring process, although you no longer get immediate displays of the values. Each time you click the mouse button you get another anchor and indication of the total distance measured from the first anchor to the last. The direction of the last line anchored is shown as "Bearing." You also get a reading for "Area."*

Under the right conditions, if an imaginary line were drawn from the last anchor to the first anchor (the starting point), the sequence of lines might define a polygon; the area indicated is the area of that polygon. If the polygon is not properly defined (that is, if any of these lines (including the imaginary last line) crosses any other), then the area number displayed is meaningless.)

{__31} *To end the measuring process, and start another set of measurements, make a final anchor with a double click.*

{__32} *Measure the distance from the leftmost side of the map window to the rightmost. It should agree approximately with the legend at the top of the map. Measure some other distances. What is the distance between the two points farthest apart on the GPS track?* _____

---

* The volcano has erupted since this data set was made; things might be different on the ground now.

{__33} *Change the Distance units to Miles and the Area units to Square Miles. Change the Not In Feature type so that the GPS points are joined. Measure and record here the approximate length of the overall GPS track. _____. What is the area of the volcano inside the track? _____.*

{__34} *Under Options ~ Units choose True North. Switch measuring off, by clicking the Select icon next to the "Measure" tool.*

Examine Properties of GPS Fixes

{__35} **Determine the location, time of acquisition, and other properties of a particular point:** *Choose "Position Properties" either from the "Tool" icons (it's a "+" sign with a "?" mark) or from the "Data" menu. A "Position Properties" window will appear. Resize the map window and move the Position Properties window so that they do not overlap. (You may have to "Zoom to Extents" so you can see the entire GPS track.)*

{__36} *Select a position anywhere on the map (but off the GPS track) clicking with the left mouse button. Note the results in the Position Properties window. Now select a position on the GPS track with the mouse. Note that the Position Properties window gives more information: You should see latitude, longitude, meters above mean sea level, the date and time the point was taken (in local time, to the nearest one-thousandth of a second), the position sequence number, and the "status" (whether it is a 2D or 3D point and whether the fix has been "corrected"). (In any of the work in this book, if you see a 2D point, it generally means an error has been made in collecting the data.)*

{__37} *By choosing the DOPs tab in the Position Properties window you can see the PRN numbers of the satellites that were used to determine the position. If the receiver had been set to record DOPs at the time the reading was taken, those values would appear as well.*

{__38} *The buttons at the top of the Position Properties window ("First," "<<," and so on) will allow you to look at the first point (First), the next point (">," not ">>"), the last point (Last), and so on. Use these tools to move the point selection (a little "x") along the path of the GPS track. In which direction was the road traversed? Clockwise ( ). Counter-clockwise ( ).*

{__39} *About what time did the trip begin? _____ What are the coordinates? _____ Record the same information for the end of the trip. _____. _____.*

{__40} *Approximating from the distance measurements you took from the display, estimate the average speed, in kilometers per hour, of the car that was carrying the antenna. _____.*

Looking at More Detail

{__41} *Change the display of Not In Feature data so that the GPS fixes are joined. (Use View ~ Layer ~ Feature.)*

{__42} *From Data ~ Position Properties, select the location of the beginning point of the trip (use First). Close the Position Properties window.*

{__43} *Use the "Zoom In" tool (a magnifying glass with a "+" on it) to point to the approximate area of the beginning of the trip and press the left mouse button. The area identified will move to the center of the screen and the distance between the points will be magnified. Notice that the scale and the distance across the window change. The distance across the window is halved; the scale is doubled. Zoom in again. Again.*

{__44} *Notice also that the track becomes more jagged as you zoom in. You can tell that it consists of a series of points connected by straight-line segments. Use the measure tool to get an idea of the size of these "jags." While the road might have followed this jagged course (it didn't—I was there) it seems more likely that the deviations are due to error in successive readings. I discuss these errors in considerable detail in Chapter 4.*

{__45} *Now de-magnify the area using the "Zoom Out" tool (select it from the "View" menu). Which icon is equivalent to this menu choice? _____ You can Zoom Out by simply clicking in the Map window. Zoom out again. Again.*

{__46} *From the "View" menu, choose "Zoom ~ Extents" to return to the original GPS track display.*

{__47} *Repeat the above "zooming" steps, but instead of simply clicking a point, drag a box around the area of interest. When zooming in, map features inside the box you drag are enlarged to fit into the map window. When zooming out, the box you drag suggests the box that you want the entire currently visible portion of the map to fit into. The smaller you make this box, the greater the degree of "zooming out." The map features shown before the zoom out will fill this box; those features will be centered at the center of the box. Finish this step by Zooming to Extents.*

The file B021602C represents the first part of the drive back from the volcano.* The trip back also includes the files B021603A and B021604A.

---

* Note the name of the file *from* the volcano. The name implies that the file data were collected on the next day (February 16). Well, it **had become** the next day in Greenwich, England. (Greenwich (pronounced *Gren-itch*) is near London.) In Hawaii this all took place in one afternoon. (Not only was the day different on the Big Island, the weather was better.) The point here is that these file names are based on UTC time, not local time.

{__48} *From the same window from which you opened VOLCANO.SSF, open B021602C.SSF while also retaining the existing (VOLCANO) file:* There is a procedure here. You may open several files at the same time by selecting the first one with a simple left mouse click on the icon. Add successive files by holding down the "Ctrl" key and clicking on the file name icons that you want to add. (You can also "de-select" file names by holding down the control key and clicking on a file name icon.) Once you have the set of files you want opened (two files, in this case) press Open and the window will disappear, leaving you with a recomputed map.

{__49} Get the "Open" window again by clicking on an icon. Now add files B021603A.SSF and B021604A.SSF. Another trick: if you hold down the Ctrl key before you begin selecting files, those files currently open, indicated by shading, will also be selected as you select new files.

{__50} Now open just the VOLCANO.SSF file by itself and display it.

Manipulating Files

{__51} *Select the three files leading away from the volcano to prepare for combining them into a single file:* Under the menu "Utilities" select "Combine." In the resulting "Combine Data Files" window, press "Browse" to bring up a list of files in the folder. Note that as you highlight a file name by (single) clicking on it you get information on that file. Pick the three files from the list, using the same approach as you did when you opened them originally. (The little input line under "File Name" may not be large enough to contain all the names, but if they are highlighted in the list they are set to be combined.) Click "Open" to send the file names to the "Combine Data Files" window.

{__52} *Combine the three files into a single file named "AWAY.SSF":* Press "Output File." Navigate to ___:\GPS2GIS\HAWAII, so that the "Save in" folder field says "Hawaii." Under "File Name" type "AWAY.SSF." Press the Save button. (You may have noticed by this time that file names are not case sensitive. I use caps here in the text for better contrast with the other words in the sentence.) Back in the Combine Data Files window, make certain that the "Sort" section indicates that the files should be combined chronologically (in time order) and, after checking all the parameters, okay the "Combine Data Files" window. Review the information (it should appear as in Figure 3.4) and then click OK. If you are told the file AWAY.SSF already exists, elect to overwrite it.

{__53} Return to "File ~ Open" and open the two files VOLCANO.SSF and AWAY. SSF. You should see the trip around the volcano and the return trip. Open the Position Properties window and select points along each route—noting the file names.

The trip *to* the volcano was along the same two-lane road as the trip away from it. The "to" trip was recorded as B021520B.SSF.

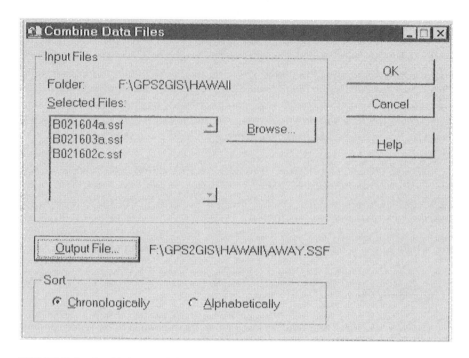

FIGURE 3.4   Combining multiple data files.

{___54} *Open VOLCANO.SSF, AWAY.SSF, and B021520B.SSF. Note that it is hard to distinguish between AWAY.SSF and B021520B.SSF.*

Pathfinder Office allows you to represent different feature types with different symbols. But it doesn't allow you to represent different files of the same feature type with different symbols. Since all three of the displayed files are of feature type "Not In Feature" they are all displayed in the same way. I would like you to see the B021520B file and the AWAY file with different symbols so that you can contrast them. I propose a ruse. Usually, different feature types are generated with a data dictionary, of which more later. But the software allows three feature types to be made from Not In Feature data. They are Point_generic, Line_generic, and Area_generic. The ruse is that we are going to convert the B021520B file into a Line_generic file so we can display it differently. We will use the Grouping Utility, normally used to group a number of files together, to convert the B021520B file.

{___55} *Use the "Grouping" utility to take the Not In Feature data of B021520B. SSF and convert it into the "Line_Generic" feature type: Under "Utilities" select "Grouping." (Or you could press the Grouping icon on the vertical Utility bar.) On the Grouping window in the Selected Files section, click "Browse," navigate to and select the B021520B file. Click Open. For an Output File, navigate to the ___:\GPS2GIS\HAWAII folder. Reject the suggested name "grouped.xxx" and use TOWARDS.SSF instead. The grouping method should be "One group per input file" and the feature type you create should be "Lines." Read the "Tip" at the bottom*

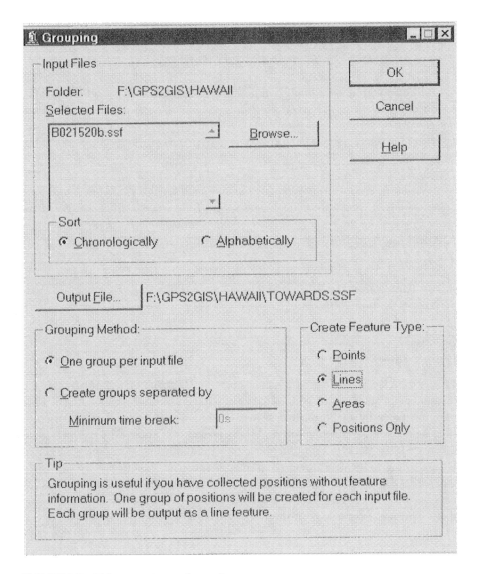

FIGURE 3.5   Using grouping to change feature type.

of the page, then click OK and the utility will finish. (If it complains that the file already exists, select the option that lets you overwrite it.) See Figure 3.5.

{__56} Open the SSF files VOLCANO, AWAY, and TOWARDS. Under View ~ Layers ~ Features change the "Not In Feature" line style to the thinnest green line. Change the "Line_generic" symbol to moderately thin unconnected black dots. Check the boxes so that all will be shown on the map. Once they are displayed, the dots from TOWARDS will overlay much of the line from AWAY. You can form an idea of the degree to which these tracks are congruent.

{__57} *Make sure the units are Meters. Pick a southern part of the road to the volcano and zoom up on it a lot (so that maybe only five or so dots of TOWARDS.SSF show). What is the approximate distance between the dots?* _____

{__58} *Zoom way up on a single dot (pick one that stands away from the line a bit) so that only it appears on the map with the green line. What is the distance between the TOWARDS.SSF dot and the AWAY.SSF line?* _____. *Since the width of this two-lane road is only about 6 meters, to what do you attribute the difference you measured?* _____ *Zoom Extents.*

{__59} *You've seen the window that will allow you to keep Not In Feature data from being displayed. Find that window and turn off the volcano track and the drive away from it. The "Big Island" has two major cities. Using a map or atlas or the Internet, determine from which city the GPS track TOWARDS the volcano originated.* _____.

{__60} *Finally, from the Data menu, turn off the Position Properties window. From the View menu turn off the Map.*

PROJECT 3B—ROOFTOP I

In this exercise you will find a single point in 3D space that approximates the location of an antenna that collected data at a fixed location. We look first at a file of points taken by a GPS receiver whose antenna was stationary over a period of time before selective availability (SA) was turned off. These GPS readings were taken from the top of a roof of a house in a Kentucky city. The idea here is that we are trying to locate a single geographic point as accurately as possible. To do that we take lots of fixes and average them. The data taken were "Not In Feature" data but we will convert them to a single Point-generic datum so we can plot the fixes. Then we take a look at three files taken at the same place after SA was turned off.

{__1} *Start the Pathfinder Office software if it is not already running.*

{__2} *Make a new project named ROOF_yis, where "yis" represents your initials. Modify the "Project" so that it references the folder ___:\GPS2GIS\ROOFTOP. (You will be told that the folder already exists and asked if you want to continue anyway. You do.)*

{__3} *Prepare to open the file B022721A.SSF which you will find in the folder ___:\ GPS2GIS\ROOFTOP. (The right folder, with the right file name, should come up since your active project is ROOF_yis and it references the ROOFTOP folder.) Write the day and month of the file, as indicated by the file name (_____) and then open it—but don't map it yet.*

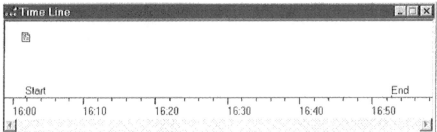

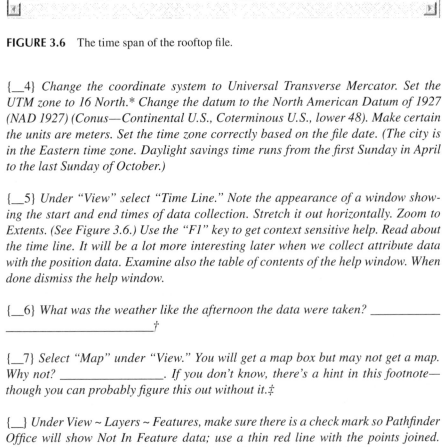

FIGURE 3.6 The time span of the rooftop file.

{__4} *Change the coordinate system to Universal Transverse Mercator. Set the UTM zone to 16 North.* Change the datum to the North American Datum of 1927 (NAD 1927) (Conus—Continental U.S., Coterminous U.S., lower 48). Make certain the units are meters. Set the time zone correctly based on the file date. (The city is in the Eastern time zone. Daylight savings time runs from the first Sunday in April to the last Sunday of October.)*

{__5} *Under "View" select "Time Line." Note the appearance of a window showing the start and end times of data collection. Stretch it out horizontally. Zoom to Extents. (See Figure 3.6.) Use the "F1" key to get context sensitive help. Read about the time line. It will be a lot more interesting later when we collect attribute data with the position data. Examine also the table of contents of the help window. When done dismiss the help window.*

{__6} *What was the weather like the afternoon the data were taken? _____ _____†*

{__7} *Select "Map" under "View." You will get a map box but may not get a map. Why not? _____. If you don't know, there's a hint in this footnote— though you can probably figure this out without it.‡*

{__} *Under View ~ Layers ~ Features, make sure there is a check mark so Pathfinder Office will show Not In Feature data; use a thin red line with the points joined. Apply Zoom Extents to both the map (after making it active) and the Time Line (after making it active).*

{__8} *Now that you can see all the fixes that have been taken, what is the approximate distance between the fixes farthest apart? _____*

---

* I could have asked you to calculate the UTM zone from the latitude and longitude. You will have to do that in the future. For the formulas to calculate the correct UTM zone for a latitude and longitude pair, see Exercise 3-4.

† Just checking to see if you were paying attention. Interestingly enough, we could have attached other devices to a GPS receiver that would have let us know a lot about the weather. More about that later.

‡ In the last project I had you turn off Not In Feature data in Feature Layers.

{__9} *For the most part, the time-adjacent points you see here are close together in terms of location. The GPS track meanders around the space sort of casually. But note that toward the middle of the window some jagged spikes appear. Zoom up on them. With the selection tool, select a point at the end of one of the spikes.*

{__10} *Open the Position Properties window and select the DOPs tab. Use the mouse-driven selection tool and the > and < keys to move along the GPS plot. Note that a point at one end of a spike is computed by using one set of four satellites, while a point at the other end uses a different set.*

By using Position Properties and moving from point to point with > and <, you were able to see that some time-adjacent points were recorded in widely different positions. What is going on here is that the "best PDOP" constellation switches from one set of satellites to another and then changes back again. This can happen if a satellite's signal is blocked by the operator's body or some other opaque entity—or if the process of calculating PDOP gives alternating results when the values are very nearly the same. Such spikes make the plot ugly and cause the receiver to use some additional memory for recording the satellites in the new constellation. But in Chapter 4 you will see that there is a way to eliminate much of the variation of the fixes from the true position value.

{__11} *In Position Properties click the Summary tab. How many points are there? _____. Over what period of time was the file collected? (Click First and Last; also see the Time Line.) _____. What UTM Zone contains the points? (Check under Options, Coordinate System, presuming you set it correctly.) _____. What city contains the points? (You might have to change the coordinate system to latitude and longitude and then go to an atlas or a World Wide Web site _____.)*

Make a Single Point That Represents the Fixes

We previously used the Grouping utility to convert raw (positions only) GPS fixes into a Line_generic feature—recall converting the GPS track toward the volcano in Project 3A. This time we will convert a set of raw GPS fixes into a single point, of feature type Point_generic. In general, Grouping allows you to use one or more input files to create output file(s) with features. We will use it to average a set of points in a file (that is, make a single point such that the *x*-coordinate is the arithmetic mean of all the *x* values, same for *y*, same for altitude (*z*)). The feature type Point-generic allows us to pick a symbol for mapping the point.

{__12} ***Make an SSF file that consists of a single point that represents the centroid (the three averages) of all the points in the file:*** *Under Utilities select Grouping (or pick it off the Utilities Toolbar). For the input we will use only the 900-point file B022721A.SSF in the GPS2GIS\ROOFTOP folder. Put the output into the ROOF_ yis Project folder ___:\GPS2GIS\ROOFTOP. The software proposes that you call the output file "GROUPED" but you should make it instead "ROOFMEAN.SSF." Under Create Feature Type, select Points. Click OK. The Grouping utility will create* *a Point_generic file named ROOFMEAN.SSF consisting of a single point. The file B022721A.SSF will remain as a 900-point Not In Feature file.*

{__13} **Display ROOFMEAN.SSF in red, using a round symbol:** *Open both B022721A.SSF and ROOFMEAN.SSF. Select Point_generic from the View ~ Layers ~ Features window, then select the Symbol you want. Also set things up so you will display B022721A.SSF with a thin, light cyan line. Draw the map with both files represented.*

{__14} *Use the pointer to select the point that represents ROOFMEAN.SSF. (When it is selected a box will appear around it.) Since the ROOFMEAN.SSF file consists of a single point, what you have here is a graphical representation of the 3D centroid—the three means (averages) of the 900 Cartesian points, though of course you cannot see the altitude dimension. Use Position Properties to assess the altitude. What is it?* _____.

ROOFMEAN.SSF represents the best approximation you can make of the actual location of the antenna, if all you have to go on are the data in B022721A.SSF. In Chapter 4, which discusses differential correction, you will see how this approximation can be improved considerably.

When SA Went Away, Things Got Better—A Lot Better

Now that SA has been turned off, individual fixes are almost always closer to the true position of the antenna. To demonstrate this, you will simultaneously open three files taken in the same place as the original, but without the intentional perturbation of SA. Then you will average the fixes in these files and look at how they compare with ROOFMEAN.SSF.

{__15} *Open the files A090214A, B, and C together and display them. Measure the distance between the fixes farthest apart. _____. Compare this with the distance you measured before for B022721A.SSF. A great improvement, no? So in terms of raw fixes collected by a GPS receiver, the variance has been improved by about an order of magnitude (a factor of 10).*

{__16} *Use the Help menu to read about the Combine command (you used it before) that lets you concatenate* GPS files so that they form a single file. Then combine A090214A, B, and C together to form ROOF_NOSA.SSF.*

{__17} *Use the Grouping utility to produce a single point from the fixes of ROOF_ NOSA.SSF to produce ROOFMEAN_NOSA.SSF. Open ROOFMEAN.SSF and ROOFMEAN_NOSA.SSF and view them. How far apart are they horizontally? _____. Vertically _____.*

---

* Concatenate: a 50-cent word that means "to connect end to end." You can make the word concatenate by concatenating the text strings "con," "cat," and "enate."

A couple of things are interesting here: First is that the two points are so close together horizontally, despite the large spread of the fixes of the files. Though we don't know the actual true point we can infer that averaging works! Secondly, the altitudes are much worse than the horizontal distances. That is simply a characteristic of GPS—though usually the errors are not this great. Bear in mind that we have not yet applied the correction mechanism that we have waiting in the wings.

Why are we looking at data taken when Selective Availability was active? After all, that "feature" has been gone a long time. But that's no absolute assurance that it won't come back. This is highly unlikely, but it's reasonable to understand the history of GIS and realize that, ultimately, it is under control of the U.S. military.

PROJECT 3C—NEW CIRCLE ROAD

Here you will examine another file taken while moving the antenna along a path. You will look at using the "Pan" and "Auto-pan to Selection" features. Additionally you will get an idea of how differential correction can improve data accuracy dramatically.

Lexington, Kentucky is circumscribed by a (mostly) limited-access ring named New Circle Road. Unlike many cities, Lexington's street pattern resembles a bicycle wheel—spokes and rim—rather than a grid. A car proceeded from the south, entered the circle road heading west, and proceeded around the city, exiting at every major spoke, making a "U" turn, and reentering the circle. The primary results of this operation were a GPS file named NEWCIRCLE.SSF and a police warning regarding illegal "U" turns.

{__1} *Set up Layers in View so that Not-In-Feature files are displayed with a thin gray line. Make the Distance units Miles. Open NEWCIRCLE.SSF which you will find in ___:\GPS2GIS\CIRCLERD. The plot should look something like Figure 3.7.*

{__2} *Zoom up so that the southernmost area of the GPS trace, including the entire north-south segment, fills the window.*

{__3} **Move the image on the screen using "Pan":** *Select Pan from the View menu or the Mouse Bar icon. Place the crosshairs on a point on the image (somewhere besides the center of the window) and click. Note that the image shifts so that the point selected moves to the center of the window. If you lose the image completely, use the "Zoom to Extents" tool to recover the entire image. Then zoom up again.*

{__4} *As with zoom, you can also drag with pan active. Select a point on the GPS track and drag it to any desired new position. If you lose the image completely, use the "Zoom to Extents" tool to recover the entire image. Then zoom up again.*

{__5} *Use the "Pan" tool to slide the image northward from the southernmost point, then westward around New Circle Road to the next exit. You can see the track going up the right side of the arterial going north-north-east, and again in the right side coming back after the "U" turn. You can also imagine the exit and entrance ramps*

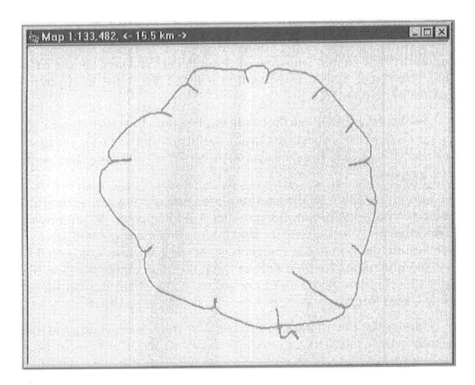

FIGURE 3.7 GPS tracks around New Circle Road.

leading from and returning to the track of the limited access road. Continue to pan around the image until you feel proficient with the panning tool. Now Zoom to Extents and, with Position Properties, select the first point.

{__6} ***Explore "Auto Pan to Selection:"*** *In the View menu, with the Map window active, put a check beside Auto Pan to Selection, or click the Auto Pan icon, to turn this feature on. Use Position Properties to select Last to place a marker on the last point of the GPS track. Now click on First (they are in about the same location; it was a complete circuit). Zoom In tightly on the selected point again. Now begin changing the selected point with the Next button (">"). Notice that when the selection would address a point that is outside the window, the selected point is moved back to the center of the window. Pan partway around the circle in this manner, perhaps with different levels of zoom. Note that you can't lose the image. Note also that it is pretty jagged in spots.*

{__7} ***Add a new file to the display—one that mimics the previous one, but whose positions have been corrected:*** * Open the files NEWCIRCLE.COR and NEWCIRCLE.SSF, by clicking on them in that order (with the Ctrl key held down*

---

* The term *corrected* is in wide use among GPS professionals. Actually, it would probably be better to say the positions have been "adjusted" to become more accurate.

to select the second one). Set up Layers in View so that Line_generic features are shown with a thin red line. (Note that the file selected first is listed last in the string of file names to be opened.)

NEWCIRCLE.COR is a file to which corrections have been applied; the process is described in detail in Chapter 4. But it is instructive for you to see the results, especially since the data you took in Project 2B may turn out to be disappointing.

{__8} *Zoom up again on the southern area and use Position Properties to select the "First" point. Be sure that Auto Pan is on. Begin moving along the trace as before. To see the improvement created by correcting the data, note the unevenness of the gray line, compared to the red one, along the first part of the path. Also notice the little detour the gray line takes just prior to the point where the car enters New Circle Road.*

{__9} *Under the File menu click Close. The map window will become blank.*

Project 3D—SA Goes Away—Background Files Help You See It

On 1 May 2000 the White House suddenly announced that selective availability (SA) would be turned off. The transition took place just after midnight Eastern Daylight Time on Tuesday 2 May 2000. Closer to home, a file was created using a receiver, with the antenna at a fixed location, from about 11:30 p.m. on May 1 until about 12:30 a.m. on May 2. The files BEFORE.SSF and AFTER.SSF were created by dividing that file with the Record Editor—a Pathfinder Office Utility program that will be described shortly. The point of division was the time at which SA was turned off.

{__1} *Make a Project named ByeByeSA_yis. Have it refer to ___:\GPS2GIS\ ByeByeSA.*

{__2} *Set the Time Zone to Eastern Daylight.*

{__3} *Starting with File ~ Open, examine the contents of the folder ByeByeSA in ___:\GPS2GIS. Open BEFORE.SSF. Note, by clicking on First and Last in the Position Properties window, that this file consists of fixes taken on May 1, approximately between 11:30 p.m. and midnight. Measure the spread of fixes. What distance is there between the farthest apart points? _____.*

You would like to contrast BEFORE.SSF with AFTER.SSF. We could open these two files together but since they are both Not In Feature files it would be hard to see them distinctly as separate. We would like to see them with different symbolization so we could compare them. As before, we could make one a Line_Generic file, but let's examine another way. Pathfinder Office allows you to display a file in the "background"—that is, the visual aspects of the file appear in the Map window, but no features in the file are selectable. We will use this "background" ability to display the file AFTER.SSF.

{__4} *Using the Units and Coordinate System choices under the Options menu, set up the proper parameters for this project: Set the Options to U.S. State Plane 1983 coordinate system, Kentucky North Zone (1601). The datum is NAD 1983 (Conus). The Distance units and Coordinate units are Survey Feet. The Time Zone remains Eastern Daylight.*

{__5} *Using View ~ Layers ~ Features makes Not In Feature data appear as a thin red line.*

{__6} *Using File ~ Background, check to see if any files are in the list of background files to load. Highlight any such file identifiers and click Remove to empty the window. Now click "Add" and navigate to ___:\GPS2GIS\ByeByeSA\AFTER.SSF and pounce on it. Okay the Load Background Files window. Under View ~ Layers ~ Background, choose a moderately thin green line to represent Not In Feature data. Once it loads you should see a green blob surrounded by red lines. Zoom up on the green area and measure the spread of fixes. _____. Quite an improvement!*

An Aerial Photo That Can Be Used as a Map: The Digital Orthophoto

A **digital orthophoto** image is an **aerial photograph,** in digital form, with a very useful special property: it has been calibrated so that it may be used as a map; the coordinates of the map are precisely specified. As you may know, most aerial photographs, even when taken with the camera pointed vertically downward, have considerable distortion, due to several factors such as uneven terrain and lens distortion. Thus, if you measure a length on a conventional photograph that corresponds to a distance on the ground, an identical length measured elsewhere on the photograph usually will not correspond to the same distance on the ground. Said concisely, a standard aerial photograph does not have a consistent scale. In a digital orthophoto image, however, objects shown on the photograph **at ground level** are related by a consistent scale, even if the terrain is dramatically uneven. (The locations of points on objects **above** ground level (for example, on the roofs of buildings) are **not,** however, shown in their correct locations.)

Digital orthophotos may come in many forms. The one we look at here is called a "**tiff**" (which stands for "**tagged image file format.**" Pathfinder Office has the capability of displaying such files as background files. The extension of tiff names is usually "**tif.**" If a tiff file is keyed to a geographical location it may be called a "**geotiff**"; geotiffs contain the parameters to anchor the image to the proper points on the Earth's surface. If a tiff image is keyed to a geographic location but contains no geographic parameters then it must be accompanied by a **world file;** world files usually end in the extension "tfw." Pathfinder Office requires that a tiff be a non-geotiff and that it be accompanied by a world file.*

---

* A geotiff may be separated into a vanilla tiff and a world file. One way to do this is with the ArcInfo CONVERTIMAGE command.

In the folder ByeByeSA are the following files:

No_sa_ncr_doq.tif—a tagged image file—the special kind of digitized aerial photograph that can be used as a map, showing the highway the automobile was traveling at the time of transition. The term **DOQ** in the name stands for digital ortho quadrangle; quadrangle is just another term for an "almost rectangular" map.*

No_sa_ncr_doq.tfw—the "world file" for No_sa_ncr_doq.tif containing numeric parameters that tell Pathfinder Office where to place the image No_sa_ncr_doq.tif.

NoSA_Moving.SSF—a subset of a file taken in a moving automobile covering the minutes of transition when SA was discontinued.

{__7} *Open ___:\GPS2GIS\ByeByeSA\NoSA_Moving.SSF. Represent the file with a medium yellow line. Set the Time Zone to Eastern Daylight. Bring up the time line and note the time span of the file. If you look at Position Properties you will note that a number of fixes have been deleted. The idea was to capture just the time surrounding the transition.*

{__8} **Prepare the map of the GPS track so that the digital orthophoto will be properly positioned:** *Before the aerial image can be displayed the geographic parameters of the map must be properly set. The DOQQQQ† (stands for digital ortho quarter quarter quarter quadrangle—since it is one sixty-fourth of a standard USGS 7.5-minute quadrangle) is in the U.S. State Plane 1983 coordinate system, Kentucky North Zone (1601). The datum is NAD 1983 (Conus). The coordinate units, altitude units, and distance units are Survey Feet. Using the Units and Coordinate System choices under the Options menu, make sure these parameters are set up correctly.*

{__9} **Bring up a digital orthophoto image as a background to the files you have selected:** *From the File menu select Background. In the Load Background Files window remove any files that show up and then click Add. The window that appears (Add Background Files) might or might not represent the folder ___:\GPS2GIS\ ByeByeSA. (If it doesn't, navigate to that folder.) Select the file no_sa_ncr_doq. tif and click on Open. Read and okay a reminder if it shows up. Back in the Load Background Files window review the System, Zone, and Datum specifications. It probably appears that all is well, but it may not be. Click on Change to examine and correct parameters so they match the ones given in the previous step. Be particularly careful regarding Coordinate Units. In Kentucky they are in Survey Feet! Okay the Coordinate System window. Okay the Load Background Files window. An image should appear in the Map window.*

---

* We say "almost rectangular" because quadrangles frequently follow lines of latitude and longitude; depending on the projection they are in the sides may not be straight.

† DOQQQQ—this is the author's name for it. You won't find this anywhere official. You will find DOQQ (Digital Ortho Quarter Quadrangle) since the USGS markets them as a quarter of a standard 7.5-minute quadrangle. But now the most usual description is DOQ, for digital ortho quadrangle.

FIGURE 3.8   Watch Selective Availability go away.

{__10} *Make the main Pathfinder Office window occupy the full screen. Make the Map window wide and not too deep, so you can have Position Properties and Time Line up also. Please see Figure 3.8. Zoom up on the part of the yellow track that crosses the DOQ. Look at the GPS tracks that represent the part of the trip on the divided highway that took place during the transition from "SA on" to "SA off." First the track, coming from the southwest, is way off the road, then you see some jagged spikes that occurred probably due to the transition, and finally the trace is at least in the right-of-way of the highway. With Position properties you can ascertain that the transition was completed at about 13 seconds after 12:01 a.m. on May 2, 2000. To see an image of the transition look at Figure 3.8.*

PROJECT 3E—YOUR DATA

Now that you have a good idea of how to use Pathfinder Office to display and analyze data taken with a GPS receiver, use this knowledge on the data you collected over a fixed point in Chapter 2. (If you don't have your own data use the ROOFTOP file B022721A.SSF and settings from PROJECT 3B.)

{__1} *Before beginning the computer work:*

- *Unless your instructor tells you otherwise, you must use the machine onto which you loaded your Chapter 2 data;*
- *Recall the name of the Project to which you transferred your data. It is probably DATA_yis (where "yis" indicates your initials). The folder in which you will work is ___:\MyGPS_yis\DATA_yis.*

- *Bring with you the full name(s) of the file(s) you collected. Also have with you the values you calculated by hand in Project 1C.*
- *Bring the detailed map(s) of the area(s) where you took the data so you can compare what appears on the screen with the features shown on the map.*

{__2} *If necessary, start the Pathfinder Office software. Select Projects in the File menu. Make sure that the Project name is "DATA_yis." Click on the Open Folder icon on the Project toolbar. Click the "Details" button to see the dates on the files and folders. Review the files that are there to be sure that you are in the right folder. You should see a Backup, Base, and Export folder as well as your files. Select, with a single click, the first of the files you took. Look at the information about that file at the bottom of the window. Close the window without opening the file.*

{__3} *Examine the map you used in Project 2A to determine the correct units, coordinate system, and datum to use. Set up the software with these parameters.*

{__4} *Under File ~ Open, select the data file you uploaded in Chapter 2 that corresponds to Project 2A. Under View ~ Layers ~ Features, set it up to display with a green line, joining the points. From the Standard toolbar, click Map and Time Line.*

{__5} *The representation now on the screen should be of the points (fixes) you collected, connected in the time order in which they were logged by the receiver.*

The following steps (6 through 15) assume that you took point data in Chapter 1 and Chapter 2 in the same place.

{__6} *In Chapter 1 you collected data by pencil and paper, averaged the values, and wrote down the results. You set the GeoExplorer rover options before that session according to:*

- *Coordinates ~ Deg & Minutes (and decimal fractional parts thereof)*
- *Datum: (set to your map)*
- *Units ~ Custom ~ Distance: (set to your map)*
- *Units ~ Custom ~ Altitude Units: (set to your map)*
- *Units ~ Custom ~ Altitude Reference: Geoid (MSL)*
- *Date & Time ~ Set Local Time: (adjust to local time)*

Set the Pathfinder Office options to the equivalents.

{__7} *Select Position Properties. Click on a fix that is toward the middle of the cluster of points. Read its location at the bottom of the screen and write it down.*

- *Latitude _____*
- *Longitude _____*
- *Altitude _____*

Is it about where you expected it to be? Compare its coordinate values against the averaged values in PROJECT 1C that you wrote down at the time.

We could make a Point_generic file (average all the fixes into a single point) as we did in Project 3B, but let's look at another way of getting the average. The data collected by the GPS receiver and downloaded into the machine take the form of **records,** which are numbered strings of text. A given GPS file will have quite a few records—at least one for every fix taken, and usually a lot more. You can look at these records with a Trimble Utility called the SSF Record Editor. When the Record Editor is invoked it automatically averages the coordinates of all the fixes and appends additional statistical records to the file at the end. In the step below you will open the file in the Record Editor, look briefly at a few records, and then read the coordinates of the "average" point.

{__8} *Go into the file menu and pick "Close" to close the current file. (The Record Editor will not calculate statistics on a file if the file is open.) You have already set up the proper parameters of datum, coordinate system, and units. In the Utilities menu select Other, then SSF Record Editor. The appropriate window will come up. Under File **in that window** click Open. You should then be able to select the file you have been working with from the window that shows up. Click Open. Make the SSF Record Editor window occupy the full screen.*

The resulting window is divided into two parts. (You can vary the position of the dividing line between them by dragging it with the mouse; "Split" in the View menu also works.) In the top part are the records of the file, beginning with a header record whose three letter designation is HDR. The HDR record is highlighted. In the lower part of the window are details in a more readable format relating to that highlighted record. Also, some records, like HDR, contain information not shown on the upper part of the window. This information appears in the bottom portion of the window. Make sure that the bottom part of the window is large enough to see all the text lines. Please see Figure 3.9.

Most SSF files have several types of records. Shortly you will see that toward the middle of the file you will find position (POS) records that tell you the time and position of each separate fix. At the end of the file are statistics records (STS).

{__9} *Look at the information in the HDR record. Then use the down arrow key and/or Page Down to look at records farther down in the file. Finally, drag the vertical slider to the bottom so that the STS records that were added to the end of the file show up. Use the "up arrow" key to select the "Latitude Stat" record. Write the coordinates here.*

- *Mean Latitude* _____
- *Mean Longitude:* _____
- *Mean Altitude:** _____

---

* At the time of this writing, previous versions of Pathfinder Office did not correctly calculate the altitude when the option "Measure Altitude from MSL" was in effect. If the altitude seems incorrect, it may be the altitude based on Height Above Ellipsoid. You could correct this by hand since you know (from a calculation in Chapter 1) the difference between HAE and MSL in your area.

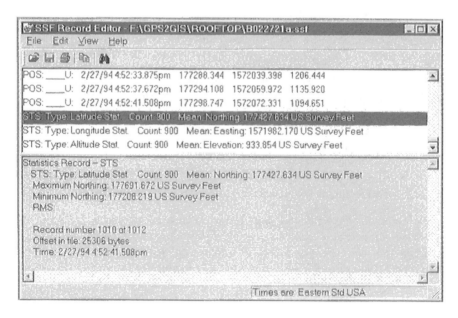

FIGURE 3.9 Finding statistics from the Record Editor.

{___10} *From the File menu exit the SSF Record Editor.*

{___11} *Some of the questions below have several blanks. Assuming you took data for both PROJECT 1B and PROJECT 2A in the same location, record the calculations you made in PROJECT 1C. Compare them with the averaged position you just found in PROJECT 2A.*

- *What is the average latitude in PROJECT 2A (from the Record Editor)? _____ (What was your Project 1 answer? _____) What is the difference in meters? _____*
- *What is the average longitude?_____ (What was your Project 1 answer? _____) What is the difference in meters? _____*
- *What is the average altitude?_____ (What was your Project 1 answer? _____) What is the difference in meters? _____*

{___12} *Compare the position given by the software with the location presented by the map. Do they agree? Answer "yes" if the horizontal measurement is within about 5 meters or so of the true location horizontally, and about 12 meters vertically. ____*

{___13} *Set up the datum, coordinate system, and units to correspond to those of the map you used in PROJECT 2B (where you moved the antenna along a path).*

{___14} *Bring in the file of PROJECT 2B and display the path you took while moving the antenna. Now query various points on the display and compare their coordinates with those on the map. Does the display correspond to the actual path you took?*

{__15} *Add the file that contains the fixes you collected in PROJECT 2A. Are the two sites in the correct relative position? If not, you may be using different projections, different datums, or different coordinate systems.*

EXERCISES

Exercise 3-1

Use the file ___:\GPS2GIS\EXERCISE_CH3\EXER3-1U.SSF to determine the general area it represents. Now add the file EXER3-1S.SSF as well. The beginning and end of these files are in the vicinities of two major universities on the North American continent. Which ones?

Exercise 3-2

Display EXER3-2H.SSF using dots not joined together. What town does the westernmost end of this path represent? (Think "liqueur.") Now open EXER3-2N.SSF. What general area is represented? Add EXER3-2S.SSF (the N and S stand for North and South). Note that you may display two files together—even those taken at different times.

    Now add EXER3-2O. (That's an alphabetic *O*.) Zoom up on this last track. What major U.S. city would you expect to find on this circuit? Change from dots unjoined to dots joined. Notice how that can confuse things.

Exercise 3-3

File EXER3-3S.SSF was generated on a sailboat that went under a bridge. A second file, EXER3-3C.SSF, was generated by a car driving over the bridge. Where did all this take place? What are the latitude and longitude coordinates of the point where the paths of the boat and car cross (use the WGS-84 datum)? What is the name of the bridge? What is the name of the body of water? What major U.S. Air Force base is nearby? When you add file EXER3-3T.SSF, what major city is indicated?

Exercise 3-4

The number of the UTM zone that contains a geographic point depends on the longitude of the point. (The longitude for a point in the western hemisphere is a negative number. That is, 100.5° West Longitude is –100.5°, so you would use this **negative number** in the formula below.)

To calculate the UTM zone for the western hemisphere: calculate

    (180° + Longitude) / 6° and round the resulting quotient up to the next integer.

To calculate the UTM zone for the eastern hemisphere: calculate

    (Longitude / 6°) + 30 and round the resulting quotient up to the next integer.

  (a) Find the UTM zone for Latitude 83.5° North, Longitude 100.5° West.

    _____

  (b) Find the UTM zone for Latitude 35.3° North, Longitude 136.5° East.

    _____

Exercise 3-5

A **waypoint** is a 3D point that has a name. Waypoints are used for navigation (for example, they may be the locations of radio-navigation antennas) or to mark the positions of objects. They may be generated directly by the Pathfinder Office Software and loaded into a Trimble receiver. Or a GPS receiver may calculate a waypoint from satellite data, or have a waypoint specified using the buttons on the front of the datalogger. However or wherever they are, a waypoint consists of a name and a location. Like other geographical data, the numerical descriptions of the position of the point may be specified in any of several coordinate systems.

The point of this exercise, besides being the briefest introduction to waypoints, is to show you what a difference the selection of a geographic projection can make in graphical presentation of data. In Pathfinder Office close any open files. Make sure the Map window is open. Set the coordinate system to Latitude and Longitude. Set the Distance Units to meters. Under View ~ Layers ~ Waypoints pick a blue diamond. Now select File ~ Waypoints ~ Open and navigate to ___:\GPS2GIS\EXERCISE_ CH3\lat_lon_square_north.wpt. Pounce on that filename. A map of four waypoints should be displayed as should a window named Waypoint Properties that names and describes the waypoints in the wpt file. (If the Waypoint Properties window doesn't open you can click an icon on the toolbar or place a check beside Waypoint Properties in the Data menu.)

Notice from the Waypoint Properties window that the four waypoints (named NW, SW, SE, and NE) form a "square" (in terms of lat-lon coordinates) whose lower left (southwest) corner is at 83°N and 101°W and upper right corner (NE) is at 84°N and 100°W. If you use the Data ~ Measure tool, you find that there is a much greater distance between the points that line up vertically than those that line up horizontally. (I deliberately picked locations in the far north to demonstrate the differences in measurements of latitude and measurements of longitude.) What is the north-south difference in meters? _____ What is the east-west difference in meters between the northernmost waypoints? _____ Further, there is a difference of about 2000 meters between the pair of northernmost waypoints and the southernmost ones. Which pair is farther apart? _____.

To see how showing a map in latitude and longitude distorted the distances, change the coordinate system to UTM where you will get a more satisfying presentation. (What you will see initially may look pretty strange—since you haven't set the UTM zone. Set it to the Zone you calculated in Exercise 3.4 for latitude 83.5° North, longitude 100.5° West.)* What you should see now is approximately a skinny trapezoid, with the top side imperceptibly shorter than the bottom side. Again you could measure the distances with the measuring tool. You should get the same values as before.

---

* Future versions of Pathfinder Office may calculate the UTM zone automatically from the latitude and longitude coordinates.

More Exercises

In the folder ___:\GPS2GIS\Exercises_and_Exams are a great many SSF files taken from locations in several parts of the world. See Appendix B for a form that you may complete. The form asks you to determine the location or track of the receiver antenna. How many points were taken? Over what time period? Would you say that the antenna was at a fixed point or moving? If moving, how fast on the average? Was there much change in the altitude? Were these changes likely due to errors (remember SA was eliminated on May 2, 2000) or to an actual change in altitude?

CHAPTER 3B: STEP-BY-STEP

> If you are using Trimble Pathfinder Office go back to Chapter 3A Step-by-Step. If you are using ESRI ArcGIS continue here.

> If you are using ESRI ArcGIS software proceed with the following material: Chapter 3B Step-by-Step. The text below assumes that you have familiarity with ArcMap. It uses ArcMap to illustrate GPS technology. The files you are using have been converted to shapefiles or geodatabase files.

PROJECT 3F—VOLCANO

We begin by looking at a file generated by a GPS receiver that circumnavigated Kilauea Caldera, the active volcano on the island of Hawaii. Before you can see or analyze the data, however, several parameters must be set up. (Right. Just as when you used the GPS receiver. It turns out that, when you do GPS for GIS, a considerable part of the activity is getting settings right. Sorry.)

{__1} *From the CD-ROM that came with the book, copy the folder IGPSwArcGIS to a hard drive (such as C:) on your computer. From now on I'll make reference to this folder as ___:\ IGPSwArcGIS, leaving you to fill in the drive letter (or perhaps the drive letter and a path) to the folder.*

{__2} *Start the ArcMap software. Close all open windows (except the main one, of course). Maximize the window.*

{__3} *In the View menu, make sure that the Standard and the Tools toolbars are on.*

{__4} *Using the "Add Data" button, add the shapefile*

___:\IGPSwArcGIS\HAWAII\Volcano.shp

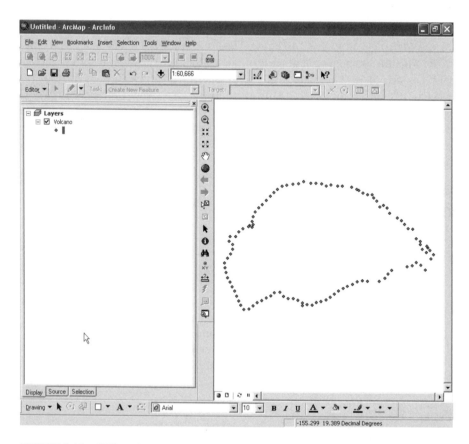

FIGURE 3.10   GPS track around Kilauea Caldera volcano.

{__5} *To change the way position data appear on the screen map, pounce* on Volcano in the Table of Contents to bring up the Layer Properties window. Under Symbology click the Symbol to see a Symbol Selector window. Make the symbol red of size four. OK. OK. Your map should look something like Figure 3.10.*

{__6} *Right click on Volcano in the Table of Contents and select Open Attribute Table from the drop-down menu. (Please see Figure 3.11.)*

{__7} *Note, from the information in the table, that the data collection start time was approximately 2:50 p.m., local time. The file is made up of 110 points (positions, fixes).*

---

* From now on, we'll call a single click on a choice, followed by the "Enter" key, a "**pounce**." You can also "pounce" with a double click but there are reasons not to do this if you aren't too familiar with Windows and/or the application software. For one, double click speeds are user selectable and a previous user may have selected a speed that makes your double click not work and leaves you wondering what happened.

FID	Shape	GPS_Time	GPS_Week	GPS_Second	GPS_Height	Latitude	Longitude	Point_ID
99	Point ZM	04:13:40pm	736	267229.906	1347.523	19.419632159	-155.243619733	100
100	Point ZM	04:13:51pm	736	267240.031	1299.556	19.419721718	-155.244973785	101
101	Point ZM	04:14:02pm	736	267251.781	1347.465	19.420908724	-155.245918860	102
102	Point ZM	04:14:13pm	736	267262.5	1375.306	19.421969870	-155.246783136	103
103	Point ZM	04:14:23pm	736	267272.938	1345.59	19.422601891	-155.248024683	104
104	Point ZM	04:14:49pm	736	267298.938	1408.114	19.423889295	-155.250186711	105
105	Point ZM	04:15:05pm	736	267314.594	1450.037	19.425307639	-155.250937269	106
106	Point ZM	04:15:14pm	736	267323.625	1388.128	19.426054532	-155.251924697	107
107	Point ZM	04:15:23pm	736	267332.375	1265.425	19.425678440	-155.252950928	108
108	Point ZM	04:15:29pm	736	267338.719	1364.03	19.427216674	-155.253809516	109
109	Point ZM	04:15:40pm	736	267349.719	1282.3	19.428074468	-155.254629222	110

Record: |◄ ◄ 0 ► ►| Show: All Selected Records (0 out of 110 Selected) Options ▼

FIGURE 3.11 Attribute table for Kilauea Caldera volcano track.

The data you are working with consists of simple position data (fixes consisting of a latitude, a longitude, an altitude). Such raw data are sometimes called "Not In Feature" data. (Positions may also be collected in such a way as to be associated with particular features, such as roads or types of vegetation, and these features may have names; I'll discuss the collection of such "feature attribute data" later in detail.)

Explore the Map

{__8} *What map scale appears on the map?* _____. *(Look in the textbox on the Standard toolbar.)*

A map scale is a fraction; it is usually much less than one (1.000); its form is "1 divided by *x*," where *x* is usually a number much larger than one. If the number *x* is one (1.0000), the map is said to have full scale. Other ways of putting it would be that the map scale is 1.0/1.0, or 1:1, or 1 to 1—all of which imply full scale. Farmers would object if you began unrolling such maps. As the scale gets smaller (nearer zero) less detail is displayed and more real estate is shown on a given size page. Many people reverse the meaning of large and small scale; they look at the denominator of the fraction, rather than the fraction itself. A 1:250 scale map is a larger scale map than a 1:500 scale map. One divided by 250 is 0.004; one over 500 is 0.002. The number 0.004 is larger than the number 0.002. Remember it this way: larger scale maps have a larger degree of detail.

{__9} *Experiment with different types of "Distance" units from the "Units" window, created by selecting from "Options ~ Units." Note the change in the value of the east-west dimension of the window (shown in the map border at the top) reflecting that different units have been specified. Try inches.* Finish up by selecting "Meters" as the distance units for the data set display.*

---

* The map scale shown is only approximate. Depending on the size of your monitor it may be off quite a bit. You can check by making the window 10 inches wide (use a ruler), multiplying the value in the denominator of the scale shown by 10, and seeing if this agrees with the distance given at the top of the window.

{__10} *With the Select Elements cursor move the arrow to the northernmost point on the map and write down the Northing (the second coordinate of the pair at the bottom of the window).* _____ *Now do the same with the southernmost point on the map.* _____ *Calculate the difference between the two values. What would you say was the north-south distance, in kilometers, of the track around the road?* _____ **

{__11} *In the "Units" window, set the "Area" to be "Hectares." (A hectare is 1/100 of a square kilometer. A hectare contains the same area as a square that is 100 meters on a side—but of course it might have any shape; there are about two and a half acres in one hectare.)*

{__12} *By right clicking the Data Frame press the General tab. Set the Display Units to Degrees Minutes Seconds. OK. This will cause the software to display latitude and longitude coordinates in the form of degrees followed by minutes, followed by seconds including decimal fractions of seconds. The other forms of display are Degrees Decimal Minutes and Decimal Degrees. Express 49°30'30.66" (49 degrees, 30 minutes, and 30.66 seconds) in these other two forms. There are 60 minutes in a degree; there are 60 seconds in a minute.†*

(1)_____ (2)_____

{__13} *Now change the Lat/Lon format to Degrees Decimal Minutes. In general, the Status bar will display the coordinates of the map position indicated by the pointer. Move the cursor around the map from south to north and observe the change in the "minutes" value shown on the Status bar. Recall that a minute of latitude is about 1845 meters. The results should agree approximately with your measurement of the distance from the north side of the road to the south. Change the style of display to Decimal Degrees.*

{__14} **Determine some data point coordinates:** *Slide the pointer down over the map. Note the status bar (at the bottom of the screen): the numbers there tell the latitude and longitude of the end of the pointer in decimal degrees. Position the pointer near the southwest portion of the "oval" (that's where the actual crater is) made by the GPS track. Write down a latitude and longitude:* _____ *If you click near a point on the GPS track, an "X" will appear, jumping to the nearest data fix. Try this. If you move the pointer off the map window the status bar will tell you the coordinate system in use. Try this.*

---

* As with many operations, there are multiple ways to achieve the same state. I won't point them all out. Once you become comfortable with the operations of Pathfinder Office you can look at the buttons and determine what shortcuts might be useful. Some of the menu options have buttons; some don't.

† To convert degrees, minutes, and seconds to decimal degrees, divide the seconds by 60, add the result to the minutes. Divide that sum by 60, then add this last quotient to the number of degrees.

A number of Internet sites allow you to display maps at various scales. At the time of this writing (and bear in mind that World Wide Web sites are constantly in flux) there are sites that allow you to put in latitude-longitude coordinates and obtain a map of the surrounding territory. Some such sites are http://www.mapquest.com, http://www.esri.com, http://www.mapblast.com, and http://www.mapsonus.com.

{__15} *If you have access to the Internet, minimize ArcMap and go to a site such as http://www.mapquest.com and find where you can get a map based on its latitude and longitude. Type in the latitude and longitude values you wrote down above (remembering that* **west** *longitudes are* **negative** *numbers). If the map-serving people have done their job correctly you should see a map of the road around the volcano. The central location indicated on the map should be approximately the location indicated on the portrayal of the GPS track. Note the similarity with the GPS track.**

{__16} *If you have access to ESRI's ArcGlobe add Volcano.shp to it as data. Zoom to the layer. Impressive, yes?*

{__17} *If you have the right software and connections you could also add the track to Google Earth. In ArcMap open ArcToolbox. Go: 3D Analyst Tools > Conversion > To KML > Layer To KML.† In the Layer drop-down menu pick Volcano. Save the file GoogleEarthVolcano.kmz in the Hawaii folder. For the Layer Output Scale put "1" and OK. Close the Layer to KML window. Now if you navigate to the file you just created and pounce on it Google Earth* **may** *open and zoom to the site.*

{__18} *Back in ArcMap with the Map window active, using the mouse, run the pointer along the icons on the Tools toolbar. Usually an explanatory box appears if you pause briefly over an icon. Find the one that allows you to measure distances on the map. (It looks like a ruler.)*

{__19} **Measure some distances:** *Click the Measure tool; a Measure window will open. If you now click any location on the map, move the cursor, and click again the Measure window will tell you the distance. If you don't like the Units (you won't because they are decimal degrees which are worthless for measurements) click the down arrow on the Measure window toolbar, pick Distance, and choose your Units—say Kilometers.*

{__20} *If you now click the left mouse button again you will anchor the moving end of the line and will see a display of the length of the line. You can continue the measuring process, although you no longer get immediate displays of the values. Each time you click the mouse button you get another anchor and distance, and indication of the total distance measured from the first anchor to the last. To end*

---

* The volcano has been erupting since this data set was taken; things might be different on the ground now.

† KML is the Keyhole Markup Language. See Wikipedia or other Internet source for a discussion.

the measuring process, and start another set of measurements, make a final anchor with a double click.

Under the right conditions, if an imaginary line were drawn from the last anchor to the first anchor (the starting point), the sequence of lines might define a polygon; the area indicated is the area of that polygon. If the polygon is not properly defined (that is, if any of these lines (including the imaginary last line) crosses any other), then the area number displayed is meaningless.)

{__21} *Measure the distance from the southernmost side of the map window to the north most. It should agree approximately with the calculations you made earlier. Measure some other distances. What is the distance between the two points farthest apart on the GPS track?* _____

{__22} *Change the Distance units to Miles. Measure and record here the approximate length of the overall GPS track.* _____.

Examine Properties of GPS Fixes

{__23} ***Determine the location, time of acquisition, and other properties of a particular point:*** *Use the Identify tool. Select a position anywhere on the map (but off the GPS track) clicking with the left mouse button. Note the results in the Identify window. Now select a position on the GPS track with the mouse. You will see the GPS date, the time the fix was taken (in local time, to the nearest one-thousandth of a second), the GPS week, the GPS second within that week, elevation in meters above mean sea level, the point's latitude and longitude, and the position sequence number—all of which came from exporting the GPS file from Pathfinder Office.*

{__24} *Open Volcano's Attribute table. In which direction was the road traversed? Clockwise ( ). Counterclockwise ( ). About what time did the trip begin? _____ What are the geographic coordinates? _____ Record the same information for the end of the trip. _____. _____.*

{__25} *Approximating from the distance measurements you took from the display, estimate the average speed, in kilometers per hour, of the car that was carrying the antenna.* _____.

Looking at More Detail

{__26} *Using the Select Features tool on the first row of the table, determine the location of the beginning point of the trip. Close the attribute table. Use the "Zoom In" tool (a magnifying glass with a "+" on it) to point to the approximate area of the beginning of the trip and press the left mouse button. The area identified will move to the center of the screen and the distance between the points will be magnified. Zoom in again. Again.*

{__27} *Notice also that the track becomes more jagged as you zoom in—it is not a straight or consistent curvilinear line. Use the measure tool, with units set to meters, to get an idea of the size of these "jags." While the road might have followed this jagged course (it didn't—I was there) it seems more likely that the deviations are due to error in successive readings. I discuss these errors in considerable detail in Chapter 4.*

{__28} *Now de-magnify the area using the "Zoom Out" tool (select it from the View menu). Which icon is equivalent to this menu choice? _____ You can Zoom Out by simply clicking in the Map window. Zoom out again. Again.*

{__29} *From the "View" menu, choose "Zoom Data > Full Extent to return to the original GPS track display.*

{__30} *Repeat the above "zooming" steps, but instead of simply clicking a point, drag a box around the area of interest. When zooming in, map features inside the box you drag are enlarged to fit into the map window. When zooming out, the box you drag suggests the box that you want the entire currently visible portion of the map to fit into. The smaller you make this box, the greater the degree of "zooming out." The map features shown before the zoom-out will fill this box; those features will be centered at the center of the box. Finish this step by Zooming to Extents.*

The trip *to* the volcano was recorded along a two-lane. This file exists in ___:\ *IGPSwArcGIS\HAWAII* under the strange name of B021520B.shp. I've included this shapefile with this counterintuitive name to illustrate how Standard Storage Format (SSF) files that come from Trimble receivers are designated, since you may run into them in your GPS work.

An SSF filename consists of eight characters:

ummddhhi

Separated for better readability, it looks like this:

u mm dd hh i

The initial character ("u" for user) of the filename is user-selectable. The middle six characters hint at the date and time: The first two digits (mm) are the month, the next two (dd) the day, and the final two (hh) the hour in UTC time. (The year the file is collected is not reflected in the file name, but it is recorded, along with many other parameters, in the file itself.)

The final character ("i" for index) is one of the 36-character sequence "A, B, C, . . . Z, 0, 1 . . . 9." Prior to recording the first file of a given hour, the "i" character is set to "A" by the receiver at the beginning of the hour. It is "incremented" to the next character in the sequence each time a file is collected and closed during that hour. This allows several files (up to 36 with the same "u" character) to be collected during a single hour, each with a unique name. If a file is open as the hour changes, **the file name does not change**. Data collection continues under the same filename. A given file may contain data taken over a several-hour span.

The name B021520B therefore indicates that this file was collected on the 15 of February in the 20th hour. The time that governs the filename is based on the time in Greenwich, England.

{__31} *From ___\IGPSwArcGIS\HAWAII add the file to the map that indicates the route taken to the volcano: B021520B.shp.*

{__32} *In ___\IGPSwArcGIS\HAWAII you will find a shapefile named AWAY.SHP. Add that to the map. This is the file of the GPS track taken while driving away from the volcano park. Note that it is hard to distinguish between AWAY.SSF and B021520B.shp since they both were recorded on the same narrow road. Zoom to full extent.*

{__33} *The "Big Island" has two major cities. Using a map or atlas or the Internet, determine which city the GPS track AWAY was headed towards. _____. Initiate a new map in ArcMap.*

You may add several files in the same folder at the same time by selecting the first one with a simple left mouse click on the icon. Add successive files by holding down the "Ctrl" key and clicking on the file name icons that you want to add. (You can also "de-select" file names by holding down the control key and clicking on a file name icon.) Once you have the set of files you want opened, press Add and the window will disappear, leaving you with a recomputed map.

{__34} *From ___\IGPSwArcGIS\HAWAII add VOLCANO.shp, Away.shp, and Towards.shp. (Towards.shp is a duplicate of B021520B.shp.)*

{__35} *Note that the dots from TOWARDS will overlay much of the line from AWAY. You can form an idea of the degree to which these tracks are congruent in the following steps.*

{__36} *Make sure the units are Meters in the Data Frame Properties window. Pick a southern part of the road to the volcano and zoom up on it a lot (so that maybe only five or so dots of TOWARDS.SSF show). What is the approximate distance between the dots? _____*

{__37} *Zoom way up on a pair of dots that have some distance between them perpendicular to the direction of the road. What is the distance between the TOWARDS. SSF dot and the AWAY.SSF dot? _____. Since the width of this two-lane road is only about 6 meters, to what do you attribute the difference you measured? _____ Zoom Extents.*

{__38} *Start ArcCatalog. Close ArcMap. Navigate to*

___*IGPSwArcGIS\HAWAII*

With the Preview tab pressed and the Preview drop-down menu indicating Geography, alternate between Volcano.shp and Volcano_UTM.shp. If you look carefully, you can see that the north-south dimension is somewhat greater with the UTM version. If you slide the cursor around on each of the maps you will alternatively see Lat/Lon coordinates and UTM coordinates.

The difference in appearance is because, when one plots latitude and longitude numbers on a Cartesian grid (which your computer screen is), the resultant figure will have a somewhat vertically squashed appearance—because a degree of longitude is not equal in distance to a degree of latitude. This is true everywhere (near the equator they are close) but as one moves north or south away from the equator the difference becomes more pronounced. At latitude 89°, a degree of longitude spans about 2 kilometers contrasted with 111 kilometers at the equator. At the Hawaiian latitude of 19° the difference is discernable, but not obvious.

ArcMap tries to make feature classes conform when presented, even if they don't agree in their underlying parameters. Thus, if you initially add a feature class which has one set of parameters and follow it with a feature class with another set of parameters, ArcMap will attempt to do "projection on the fly" so that the resultant map will appear coherent. This is both a blessing and a curse, since it may hide important differences.

{__39} *To demonstrate on-the-fly projection, start ArcMap, add data to a new map in this order: Volcano.shp, Volcano_UTM.shp. Click the check mark in front of Volcano_UTM on and off to assure yourself that the points are congruent. Since you loaded the geographic (rather than the projected) file first it takes precedence.*

{__40} *Dismiss ArcMap.*

Project 3G—Rooftop I

In this exercise you will find a single point in 3D space that approximates the location of an antenna that collected data at a fixed location. We look first at a file of points taken by a GPS receiver whose antenna was stationary over a period of time before selective availability (SA) was turned off. These GPS readings were taken from the top of a roof of a house in a Kentucky city. The idea here is that we are trying to locate a single geographic point as accurately as possible. To do that we take lots of fixes and average them. Then we take a look at three files taken at the same place after SA was turned off.

{__1} *Start the ArcMap software if it is not already running.*

{__2} *Open the file B022721A.shp which you will find in the folder ___:\IGPSwArcGIS\ROOFTOP. Write the day and month of the file, as indicated by the Trimble file name from which the shapefile was derived.* _____

{__3} *What was the weather like the afternoon the data were taken?* _____
_____*

{__4} *You can now see all the fixes that have been taken. What is the approximate distance between the fixes farthest apart?* _____ *In Chapter 4 you will see that there is a way to eliminate much of the variation of the fixes from the true position value.*

Lines connect points in the time order that they were taken. This plot looks very bad. You can see that some points taken at nearly the same time show widely different positions. What is going on here is that the "best PDOP" constellation switches from one set of four satellites to another and then changes back again. This can happen if a satellite's signal is blocked by the operator's body or some other opaque entity—or if the process of calculating PDOP gives alternating results when the values are very nearly the same. Such spikes make the plot ugly and cause the receiver to use some additional memory for recording the satellites in the new constellation.

{__5} *How many points are there?* _____. *Over what period of time was the file collected? (See the attribute table.)* _____. *What city contains the points? (You might have to go to an atlas or a World Wide Web site* _____.)

See a Single Point That Represents the Fixes

{__6} **Examine a shapefile that consists of a single point that represents the centroid (the three averages) of all the points in the file:** *In the*

IGPSwArcGIS\ROOFTOP\export

folder is a file named ROOFMEAN. This had the input of the 900-point file B022721A.SSF file. Add this to the map; enlarge it and make it a different color than the 900 points.

{__7} *Use the Identify tool to determine the coordinates of ROOFMEAN. Latitude* _____ *Longitude* _____ *Altitude* _____*feet.*

---

* Just checking to see if you were paying attention. Interestingly enough, we could have attached other devices to a GPS receiver that would have let us know a lot about the weather. More about that later.

ROOFMEAN.SSF represents the best approximation you can make of the actual location of the antenna, if all you have to go on are the data in B022721A.SSF. In Chapter 4, which discusses differential correction, you will see how this approximation can be improved considerably.

When SA Went Away, Things Got Better—A Lot Better

Now that SA has been turned off individual fixes are almost always closer to the true position of the antenna. To demonstrate this, you will simultaneously open a file taken in the same place as the original, but without the intentional perturbation of SA. Then you will look at the average of these and see how they compare with ROOFMEAN.SSF.

{__8} *In ArcMap add the file A090214B to the display. Use a symbol and color that you can distinguish from B022721A. Measure the distance between the fixes farthest apart of this second file. _____. Compare this with the distance you measured before for B022721A. A great improvement, yes? So in terms of raw fixes collected by a GPS receiver, the variance has been improved by at least an order of magnitude (a factor of 10).*

{__9} *Finally add the file ROOFMEAN_NOSA. Enlarge its symbol. Zoom in and measure the distance from ROOFMEAN to ROOFMEAN_NOSA. How far apart are they? _____.*

{__10} *Use the Identify tool to determine the coordinates of ROOFMEAN_NOSA. Latitude _____ Longitude _____ Altitude _____feet.*

A couple of things are interesting here: First is that the two average points are very close together horizontally, despite the large spread of the fixes of the files. Though we don't know the actual true point we can infer that averaging works! Secondly, the altitudes are much worse than the horizontal distances. That is simply a characteristic of GPS—though usually the errors are not this great. Bear in mind that we have not yet applied the correction mechanism that we have waiting in the wings.

Why are we looking at data taken when Selective Availability was active? After all, that "feature" has been gone a long time. But that's no absolute assurance that it won't come back. This is highly unlikely, but it's reasonable to understand the history of GIS and realize that, ultimately, it is under control of the U.S. military.

PROJECT 3H—NEW CIRCLE ROAD

Here you will examine another file taken while moving the antenna along a path. You will get an idea of how differential correction can improve data accuracy dramatically.

Lexington, Kentucky, is circumscribed by a (mostly) limited-access ring named New Circle Road. Unlike many cities, Lexington's street pattern resembles a bicycle wheel—spokes and rim—rather than a grid. A car proceeded from the south, entered the circle road heading west, and proceeded around the city, exiting at every major spoke, making a "U" turn, and reentering the circle. The primary results of this operation were a GPS file named NEWCIRCLE and a police warning regarding illegal "U" turns.

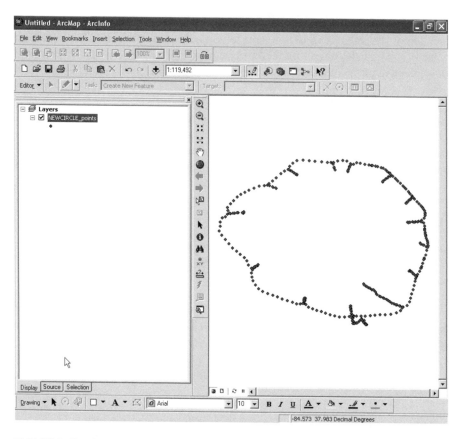

FIGURE 3.12   GPS track around New Circle Road.

{__1} *Start ArcMap. Open NEWCIRCLE_points.shp which you will find in* ___*:\IGPSwArcGIS\CIRCLERD. Make the Distance units Miles. The plot should look something like Figure 3.12.*

{__2} *Zoom in so that the southernmost area of the GPS trace, including the entire north-south segment, fills the window.*

{__3} ***Move the image on the screen using "Pan":*** *Select Pan from the Toolsmenu. Place the "hand" cursor on a point on the image (somewhere besides the center of the window) and click. Note that the image shifts so that the point selected moves to the center of the window. If you lose the image completely, use the zoom to Full Extent tool to recover the entire image. Then zoom in again.*

{__4} *You can also drag with pan active. Select a point on the GPS track and drag it to any desired new position. If you lose the image completely, use the zoom to Full Extent tool to recover the entire image. Then zoom in again.*

{__5} *Use the "Pan" tool to slide the image northward from the southernmost point, then westward around New Circle Road to the next exit. You can see the track going up the right side of the arterial going north-northeast, and again in the right side coming back after the "U" turn. You can also imagine the exit and entrance ramps leading from and returning to the track of the limited access road. Continue to pan around the image until you feel proficient with the panning tool. Now zoom to full extent, label the points with Point_ID (reminder: the Labels tab is in Feature Properties) and examine the first point with the Identify tool. What is its height? _____feet*

Frequently with GPS points we want to join them in the time order in which they were created. As you will see by adding the next shapefile, you can get a better idea of the track formed by the points.

Add the shapefile NEWCIRCLE_line from the same folder as NEWCIRCLE_points. Make the line gray (Reminder: click on the symbol; select a color.). Open the attribute table of NEWCIRCLE_line. Notice that there is only one feature: the line connecting the dots in sequence. Notice also that there are two Lengths recorded. GPS_Length is the sum of the distances between the points, and thus approximates the length of the track in two dimensions. GPS_3DLeng approximates the sum of the distances between the point as though they existed in three-dimensional space. As you would expect, the 3D length is longer. This is due to the actual changes in elevation of the road and, perhaps to a greater degree, the variance in the 3D positions caused by error. What is the 2D length of the drive? _____

Looking at the Effect of Differential Correction

{__6} **Add a new file to the display—one that mimics the previous one, but whose positions have been corrected.*** *Add the file NEWCIRCLE_corrected. Make the line red.*

NEWCIRCLE_corrected is a file to which corrections have been applied; the process is described in detail in Chapter 4. But it is instructive for you to see the results, especially since the data you took in Project 2B may turn out to be disappointing.

{__7} *Zoom up again on the southern area and begin moving along the trace as before. To see the improvement created by correcting the data, note the unevenness of the gray line, compared to the red one, along the first part of the path. Also notice the little detour the gray line takes just prior to the point where the car enters New Circle Road. Continue to pan around the track, noticing particularly at the exits where differential correction moved the line considerably away from the points.*

{__8} *Dismiss ArcMap.*

---

* The term *corrected* is in wide use among GPS professionals. Actually, it would probably be better to say the positions have been *adjusted* to become more accurate.

Project 31—SA Goes Away—Background Files Help You See It

On 1 May 2000 the White House suddenly announced that selective availability (SA) would be turned off. The transition took place just after midnight Eastern Daylight Time on Tuesday 2 May 2000. A file was created using a GPS receiver, with the antenna at a fixed location, from about 11:30 p.m. on May 1 until about 12:30 a.m. on May 2. The files BEFORE.SSF and AFTER.SSF were created. The point of division was the time at which SA was turned off.

{__1} *Start ArcMap. Add as data a shapefile named BEFORE.shp from the folder ___:\IGPSwArcGIS\ByeByeSA.*

{__2} *This file consists of fixes taken on May 1, approximately between 11:30 p.m. and midnight. Use the Measure tool, set to meters, to approximate the distance between the two points farthest apart. _____ meters*

{__3} *Now open the file AFTER.shp. This file consists of points taken just after midnight, when Selective Availability was turned off. Again use the Measure tool. The distance between the two points farthest apart in AFTER.shp is _____ .*

{__4} **Using the Units and Coordinate System choices under the Options menu, set up the proper parameters for this project:** *Set the Options to U.S. State Plane 1983 coordinate system, Kentucky North Zone (1601). The datum is NAD 1983 (Conus). The Distance units and Coordinate units are Survey Feet. The Time Zone remains Eastern Daylight.*

An Aerial Photo That Can Be Used as a Map: The Digital Orthophoto

A **digital orthophoto** image is an **aerial photograph**, in digital form, with a very useful special property: it has been calibrated so that it may be used as a map; the coordinates of the map are precisely specified. As you may know, most aerial photographs, even when taken with the camera pointed vertically downward, have considerable distortion, due to several factors such as uneven terrain and lens distortion. Thus, if you measure a length on a conventional photograph that corresponds to a distance on the ground, an identical length measured elsewhere on the photograph usually will not correspond to the same distance on the ground. Said concisely, a standard aerial photograph does not have a consistent scale. In a digital orthophoto image, however, objects shown on the photograph **at ground level** are related by a consistent scale, even if the terrain is dramatically uneven. (The locations of points on objects **above** ground level (for example, on the roofs of buildings) are *not*, however, shown in their correct locations.)

Digital orthophotos may come in many forms. The one we look at here is called a "**tiff**" (which stands for "**tagged image file format**." ArcMap has the capability of displaying such files as background files. The extension of tiff names is usually "**tif**." If a tiff file is keyed to a geographical location it may be called a "**geotiff**"; geotiffs contain the parameters to anchor the image to the proper points on the Earth's surface. If a tiff image is keyed to a geographic location but contains no geographic parameters then it must be accompanied by a **world file;** world files usually end in the extension "tfw." ArcMap will accept either a geotiff or a regular tiff accompanied by a world file.*

In the folder ByeByeSA are the following files:

No_sa_ncr_doq.tif—a tagged image file—the special kind of digitized aerial photograph that can be used as a map, showing the highway the automobile was traveling at the time of transition. The term DOQ in the name stands for digital ortho quadrangle; quadrangle is just another term for an "almost rectangular" map.† This particular image could be called a DOQQQQ‡ (stands for digital ortho quarter quarter quarter quadrangle—since it is one sixty-fourth of a standard USGS 7.5-minute quadrangle). It is in the U.S. State Plane 1983 coordinate system, Kentucky North Zone (1601). The datum is NAD 1983 (Conus). The coordinate units, altitude units, and distance units are Survey Feet.

No_sa_ncr_doq.tfw—the "world file" for No_sa_ncr_doq.tif containing numeric parameters that tell ArcMap where to place the image No_sa_ncr_doq.tif.

NoSA_Moving.shp—a subset of a file taken in a moving automobile covering the minutes of transition when SA was discontinued. It has been projected into Kentucky North Zone coordinates, so it will match with the DOQ. Before the aerial image can be displayed the geographic parameters of the map must be properly set. The failure to make sure that all features added to a map is a major cause of getting wrong answers from a GIS. One has to be particularly careful regarding Coordinate Units. In Kentucky they are in Survey Feet! Other states use International Feet. Still others use Meters.

{__5} *Start a new map in ArcMap. Add as data NoSA_Moving.shp. Represent the file with a width 2 yellow line. There aren't very many fixes here. The idea was to capture just the time surrounding the transition. An image should appear in the Map window.*

{__6} *Look at the GPS tracks that represent the part of the trip on the divided highway that took place during the transition from "SA on" to "SA off." First the track, coming from the west, is way off the road, then you see some jagged spikes that occurred probably due to the transition, and finally the trace is at least in the right-of-way of the highway. Use the Identify tool to determine when the track began. _____ a.m. The transition was completed at about 13 seconds after 12:01 a.m. on May 2, 2000. To see an image in the text of the transition look at Figure 3.8.*

---

* A geotiff may be separated into a vanilla tiff and a world file. One way to do this is with the ArcInfo CONVERTIMAGE command.

† We say "almost rectangular" because quadrangles frequently follow lines of latitude and longitude; depending on the projection they are in the sides may not be straight.

‡ DOQQQQ—this is the author's name for it. You won't find this anywhere official. You will find DOQQ (Digital Ortho Quarter Quadrangle) since the USGS markets them as a quarter of a standard 7.5-minute quadrangle. But now the most usual description is DOQ, for digital orthoquadrangle.

Project 3J—Your Data

Now that you have a good idea of how to use ArcMap to display and analyze data taken with a GPS receiver, use this knowledge on the data you collected over a fixed point in Chapter 2. (If you don't have your own data use the file A090214B_points_only.shp and settings from PROJECT 3B.)

{__1} *Before beginning the computer work:*

- *Unless your instructor tells you otherwise, you must use the network or machine onto which you loaded your Chapter 2 data;*
- *Recall the name of the Project to which you transferred your data. It is probably DATA_yis (where "yis" indicates your initials). The folder in which you will work is ___:\MyGPS_yis\DATA_yis.*
- *Bring with you the full name(s) of the file(s) you collected. Also have with you the values you calculated by hand in Project 1C.*
- *Bring the detailed map(s) of the area(s) where you took the data so you can compare what appears on the screen with the features shown on the map.*

{__2} *Start the ArcMap software and start with an empty map. Navigate to DATA_ yis that is inside the folder MyGPS_yis. Select the QuickProject shapefile that you uploaded in Chapter 2 that contains the data you took at a stationary point (Project 2A). (If you can't use your own data, use A090214B_points_only.shp from IGPSwArcGIS\ROOFTOP).*

{__3} *The representation now on the screen should be of the points (fixes) you collected.*

The following ___ steps assume that you took point data in Chapter 1 and Chapter 2 in the same place.

{__4} *In ArcMap we need to increase the number of decimal places shown at the cursor location, because the default of three doesn't give enough specificity. Go: Tools > Options > Data View and Round Coordinates to six decimal places. Apply. OK.*

{__5} *In Chapter 1 you collected data by pencil and paper, averaged the values, and wrote down the results. They were: Latitude _____ Longitude _____*

{__6} *Select Position Properties. Click on a fix that is toward the middle of the cluster of points. Read its location at the bottom of the screen and write it down.*

- *Latitude _____*
- *Longitude _____*

Is it about where you expected it to be? Compare its coordinate values against the averaged values in PROJECT 1C that you wrote down above.

EXERCISES

Exercise 3-1

Use the file ___:\IGPSwArcGIS\EXERCISE_CH3\EXER3-1U.shp to determine the general area it represents. Now add the file EXER3-1S.shp as well. The beginning and end of these files are in the vicinities of two major universities on the North American continent. Which ones?

Exercise 3-2

Display EXER3-2H.shp. What town does the westernmost end of this path represent? (Think "liqueur.") Now open EXER3-2N.shp. What general area is represented? Add EXER3-2S.shp (the N and S stand for North and South).

   Now add EXER3-2O. (That's an alphabetic O.) Zoom up on this last track. What major U.S. city would you expect to find on this circuit?

Exercise 3-3

File EXER3-3S.shp was generated on a sailboat that went under a bridge. A second file, EXER3-3C.shp, was generated by a car driving over the bridge. Where did all this take place? What are the latitude and longitude coordinates of the point where the paths of the boat and car cross (use the WGS-84 datum)? What is the name of the bridge? What is the name of the body of water? What major U.S. Air Force base is nearby? When you add file EXER3-3T.shp, what major city is indicated?

Exercise 3-4

The number of the UTM zone that contains a geographic point depends on the longitude of the point. (The longitude for a point in the western hemisphere is a negative number. That is, 100.5° West Longitude is –100.5°, so you would use this **negative number** in the formula below.)

To calculate the UTM zone for the western hemisphere: calculate

   (180° + Longitude) / 6° and round the resulting quotient up to the next integer.

To calculate the UTM zone for the eastern hemisphere: calculate

   (Longitude / 6°) + 30 and round the resulting quotient up to the next integer.

   (a) Find the UTM zone for Latitude 83.5° North, Longitude 100.5° West.

   _____

   (b) Find the UTM zone for Latitude 35.3° North, Longitude 136.5° East.

   _____

MORE EXERCISES

In the folder ___:\IGPSwArcGIS\Exercises_and_Exams are a great many shape-files taken from locations in several parts of the world. See Appendix B for a form which you may complete. The form asks you to determine the location or track of the receiver antenna. How many points were taken? Over what time period? Would you say that the antenna was at a fixed point or moving? If moving, how fast on the average? Was there much change in the altitude? Were these changes likely due to errors (remember SA was eliminated on May 2, 2000) or to an actual change in altitude?

4 Differential Correction, DOQs, and ESRI Data

In which we take a closer look at the subject of GPS accuracy and explore techniques that reduce errors.

OVERVIEW

GPS Accuracy in General

When you record a single position with a good GPS receiver, the position recorded will be probably within 5 to 15 meters horizontally of the true location of the antenna.

When a surveyor uses good, survey-grade GPS equipment, he or she can locate a point to within a centimeter of its true horizontal position. What are the factors that allow the surveyor to be 1,000 or so times more accurate than you are? This is a complicated subject. The answer includes "very good equipment," "measuring the actual number of waves in the carrier" (as differentiated from interpreting the codes impressed on the carrier), and "spending a lot of time" at each site.* We can cover only the basics in a book of this scope. But you will learn how to reduce errors so that you can record a fix to within half a meter to 3 meters of its true location. One primary method of gaining such accuracy is called "differential correction."

Differential Correction in Summary

In a nutshell, the differential correction process consists of setting a GPS receiver (called a **base station**) at a precisely known geographic point. Because the base station knows exactly where its antenna is, it can analyze and record errors in the GPS signals it receives—signals that try to tell it that it is somewhere else. That is, the base station knows the truth, so it can assess the lies being told to it by the GPS signals. These signal errors will be almost equivalent to the signal errors affecting other GPS receivers in the local area, so the accuracy of locations calculated by those other receivers may be improved, dramatically, by information supplied by the base station.

Thinking about Error

For the logging of a given point, define *error* as the distance between what your GPS receiver **records as the position** of the antenna and the **true position** of the antenna.

---

* A sarcastic surveyor of the author's acquaintance said that GPS for surveyors stood for Get Paid for Sitting.

It is useful to dissect the idea of *error*. We can speak of error in a horizontal plane and differentiate it from the vertical error. This is important in GPS, because the geometry of the satellites dictates that no matter what we do, vertical error will almost always exceed horizontal error on or near the surface of the Earth. The fact that all the satellites are necessarily above the fix being taken generally means that vertical error will be 1.5 to 2.5 as great as horizontal error.

Another useful distinction is between what we might call **random** error and **systematic** error, or bias. Random errors are deviations from a "true" value that follow no predictable pattern. Systematic errors do follow a predictable pattern. An example will be illustrative. Suppose we have a machine designed to hurl tennis balls so that they land a certain distance away on a small target painted on the ground. Of course, none of the balls will hit the center of the target exactly; there will always be some error.

What factors might cause errors? The balls are each of slightly different weight; they are not symmetrical and will be loaded into the machine in different orientations. Because it is hard to determine the effects of these factors on the accuracy of the process, we say the factors induce random errors. If there are only random errors in the process, some balls will hit short of the center of the target, some beyond it, some left, some right, and so on. If we shoot 100 balls from the machine, we will see a pattern of strikes in the area of target which appears somewhat random but which clusters around the target.

Now suppose that we had set up our machine and its target when there was no wind, but then a constant breeze of 10 miles per hour began blowing from the right across the path of flight of the tennis balls. This would create a systematic error: each ball would land somewhat to the left of where it would have landed in the no-wind condition. We will still see a random pattern of hits, but the average of all hits will be somewhat to the left of the target. This is systematic error; the "system," including the wind, causes it. To correct, we could aim the machine somewhat to the right.

Other examples of factors that might contribute to errors are the following: as the temperature changes, the characteristics of the machine may change; the atmospheric pressure and the relative humidity of the air will affect the drag on a ball; and so on. Whether these might be random errors or systematic ones might be hard to determine.

Generally, random errors are those caused by factors we cannot measure or control; systematic errors are those we can account for, measure, and, perhaps, correct for.

First Line of Defense against Error: Averaging

When I implied that the surveyor could be 1,000 times more accurate than the average person with a GPS receiver, I was being somewhat disingenuous, mostly for effect. I was comparing a single reading with inexpensive equipment with the average of many readings from expensive equipment. This is not a fair contrast, because you can improve the accuracy of the less expensive equipment by taking many readings at a fixed point. You recall that the strikes of the tennis balls, with no wind, tended to cluster around the target. GPS readings tend to cluster around the true location. We can use the fact that large numbers of random errors tend to be self-canceling. That is, the average position (if you take the means of many latitudes, of many longitudes, of many altitudes) will be much closer to the true value than the typical single measurement.

One measure of accuracy of GPS fixes is called Circular Error Probable (CEP). It is the radius of a circle expressed in a linear unit, such as meters. For a given situation, 50% of the fixes will fall within the circle, and 50% outside. Another measure of accuracy is based on two standard deviations of a normal distribution—called 2dRMS where RMS means root-mean-square. Ninety-five percent of the fixes will lie within a circle with this radius.

For NAVSTAR GPS, a number of experiments suggest that 50% of the latitude and longitude fixes you obtain with a single receiver operating by itself (i.e., autonomously) will lie within 12 meters of the true point. Fifty percent of the altitude fixes will lie within 21 meters. The 2dRMS radius is 30 meters horizontally, and 52 meters vertically. These numbers assume that selective availability is off.

In general, the more fixes you take and the more time you spend, the better your average will be. If you are prepared to take data at one point for several weeks to several months your error will get down to approximately 1 to 2 meters, basically due to the law of large numbers.* This will not be a practical way to reduce error in most applications.

Another approach, which is related to averaging in a different way, is to use "overdetermined" position finding. As you know, four satellites are required for a 3D fix. But suppose your receiver has access to five or more at a given time. Each set of four of the satellites available will provide a different opinion on the position of the point being sought. A compromise agreement based on all the satellite's input is probably better than the position indicated by any one set of four. The GPS receiver may be set to collect data in this way.

SOURCES OF GPS ERROR

Now look at the specific sources of errors in GPS measurements. Typical error sources and values for receivers of the Pathfinder class are as follows:

satellite clocks	< 1 meter
ephemeris error	< 1 meter
receiver error	< 2 meters
ionospheric	< 2 meters
tropospheric	< 2 meters
multipath	(depends on the situation and the receiver; may be large)

These values correspond to averages of many readings rather than the error that might be expected from a single reading. Although experimentation shows that the more fixes you record the better the data become, the increase in accuracy created by taking a large number of fixes really cannot be counted upon. The existence of

---

* According to Chuck Gilbert, of Trimble Navigation, writing in the February 1995 issue of *Earth Observation Magazine.*

systematic errors that might be present because of particular atmospheric conditions (or, for certain, when selective availability was active) makes the law of large numbers inapplicable. As you will see shortly, there are better ways to get really accurate data. And more accurate data is almost always required when using GPS for GIS purposes.

CLOCK ERRORS

As you know, the ability of a GPS receiver to determine a fix depends on its ability to determine how long it takes a signal to get from the satellite to the receiver antenna. This requires that the atomic clocks in the satellite be synchronized. Even a small amount of difference in the clocks can make a huge difference in the distance measurements, because the GPS signal travels at about 300,000,000 meters per second.

EPHEMERIS ERRORS

The receiver expects each satellite to be at a certain place at a particular given time. Every hour or so, in its data message, the satellite tells the receiver where it is predicted to be at time "t" hence. If this ephemeris prediction is incorrect—the satellite isn't where it is expected to be, even by just a meter or two—then the measurement of the range from the receiver antenna to the satellite will be incorrect.

RECEIVER ERRORS

These errors result from a number of factors related to receiver design, cost, and quality. Some receivers, for example, cannot exactly measure and compute the distance to each satellite simultaneously. If a GPS receiver can track up to six satellites at a single moment, then it is a six-channel receiver. In any receiver the computer must work with a fixed number of digits and is therefore subject to calculation errors. The fact is, perfection in position calculation by computer simply is not possible, because computers cannot do arithmetic on fractions exactly. (It is true that other computer operations, such as addition of integers, can be perfect.)

ATMOSPHERIC ERRORS

For most of its trip from the satellite to the receiver antenna, the GPS signal enjoys a trip through the virtual vacuum of "empty space." Half of the mass of the Earth's atmosphere is within 3.5 miles of sea level. Virtually all of it is within 100 miles of the surface. So the signal gets to go the speed limit for electromagnetic radiation for more than 12,000 of the more than 12,600 miles of the trip. When it gets to the Earth's atmosphere, however, the speed drops very slightly—by an amount that varies somewhat randomly. Of course, since the calculation of the range to the satellite depends on the speed of the signal, a change in signal speed implies an error in distance, which produces an error in position finding.

Significant changes in signal speed occur throughout the atmosphere, but the primary contributions to error come from the ionosphere, which contains charged particles under the influence of the Earth's magnetic field, and from the tropo-sphere—that dense part of the atmosphere that we breathe, that rains on us, and that demonstrates large variations in pressure and depth.

More sophisticated GPS equipment can "calculate out" most of the ionospheric error because it considers both the frequencies transmitted by each satellite. Since the ionosphere affects the different frequencies differently a correction can be calculated. Tropospheric errors, however, we seem to be pretty much stuck with, especially using the moderately inexpensive code-based equipment available to civilians. This will change when a second civilian signal is added to the system.

Multipath Errors

As indicated in Chapter 2, substantial errors may occur if a given radio signal follows two paths to the receiver antenna. This can happen if a part of the signal bounces off an object, such as a building. The arrival of two or more parts of the signal at different times can confuse the receiver and produce a false reading. Many receivers are programmed to disregard the second signal. But a problem can occur if the direct signal is blocked but the related bounced signal is seen by the antenna and recorded.

Selective Availability—A Former (We Trust) Source of Error

Until May 2, 2000, the lion's share of the "Error Budget" came from deliberate cor-ruption of the signal by the U.S. DoD. As I have said, they didn't want the civilian GPS receiver to be able to pinpoint its position. In fact, in the early days, they didn't want the civilian world to know GPS existed. Selective availability was one tech-nique the military could use to keep the system from being too accurate. (Another is called "anti-spoofing.") The errors were probably induced by dithering (making inaccurate) the signal that tells the ground receiver the exact time or by broadcasting a slightly false ephemeris. Obviously, the military didn't want to say too much about its techniques for making the good GPS signals "selectively available"—i.e., avail-able only to military receivers.

The DoD had, on occasion, turned selective availability off. (Two notable times were during the first war with Iraq and during the almost-invasion of Haiti.)

With SA off an autonomous receiver can supply much more accurate posi-tion information. The 50% measure (CEP) measure is about 12 meters horizon-tally and 21 meters vertically, due to the errors previously discussed. Ninety-five percent (2dRMS) of the fixes fall within 30 meters horizontally, and 52 meters vertically.

Reducing Errors

As it turns out, most of these errors can be "calculated out" of the measurements by the process called differential correction. To understand how it works, let me first show you the results of an experiment.

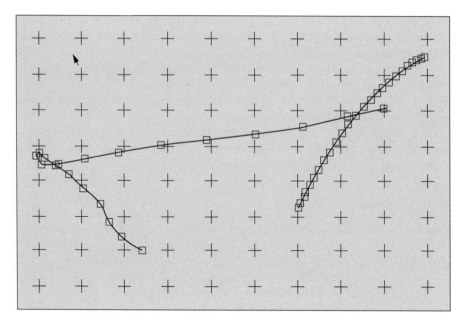

FIGURE 4.1   Two GPS files representing the same point, taken with receiver #1.

Figure 4.1 shows two sequences of points (two different files) taken with a GPS receiver placed at a single location, operating during two different time periods.*

The neat string at right was made of fixes taken about 6 seconds apart; the mess at the left was measured by taking points every 20 seconds during a different time period. (The tick marks are 10 meters apart.) You should realize that each and every fix here is attempting to approximate the same true position. They show up in different positions because of the errors discussed above. We don't know what the true position of the antenna was, but almost certainly each point misses the exact, true position by some amount.

Figure 4.2 shows data collected at the same times as in the previous figure; the computations of location were based on the same set of four satellites. The antenna for this receiver (#2) was in virtually the same geographic location as the other.

Now look at a composite of the two sets of data as shown in Figure 4.3. The points of #1 are marked with boxes; the data from #2 are marked with circles.

What is interesting is that the two patterns of fixes show remarkable similarity. Pairs of fixes—taken at the same time from the same set of satellites—appear in about the same position. This suggests that the *error* for each associated pair of points is almost identical, even though the fixes have come from different receivers and different antennas. And this suggests something which turns out to be true, though I have not proved it here: that receivers calculating from the same signals will

---

* These files were taken when selective availability (SA) was active. Files taken with SA turned off would reveal similar patterns but not as dramatically, since the amount of error has been greatly reduced.

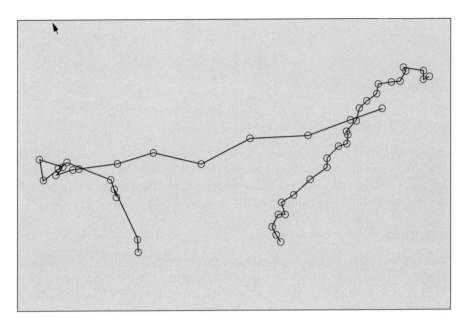

FIGURE 4.2   GPS files representing the same point as the figure above, but taken with receiver #2.

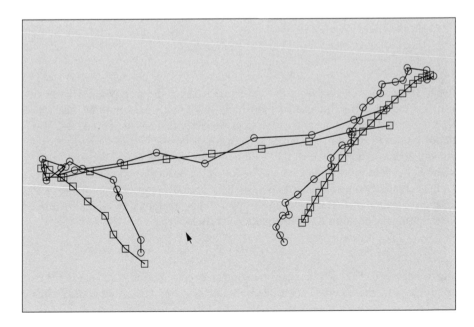

FIGURE 4.3   Composite of the representations of the two preceding figures.

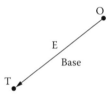

FIGURE 4.4   Error vector—observed point to true point.

suffer from almost the same errors, provided that the antennas are "close." What's close? Two receivers within 500 kilometers (300 miles) will tend to show the same magnitude and direction of errors with respect to the true locations of their antennas, provided the positions are found using the same set of satellites.

MORE FORMALLY*

The experiment described above demonstrates that much of the error is inherent in the signals—that is, the errors occur before the signals reach the receiver antenna.

To see how that helps us remove most of the error from a GPS fix, let's focus on both a single point on Earth's surface (a true point, T), and its representation in the GPS receiver (the measured, or observed, point, O).

Suppose we take a GPS receiver antenna, and place it precisely at that *known point T*—a point that has been surveyed by exacting means and whose true position is known to within a centimeter. We call such an antenna-receiver configuration **GPS base station.** Now consider that an observation O is taken by a base station receiver. So we have three entities to consider, as shown in Figure 4.4:

- the position of T
- the reading O, and
- the "difference" between O and T.

We've drawn an arrow from the measured point to the true point. This arrow, which is shown in two dimensions but which would really exist in three, has both a length (called a magnitude) and a direction. An entity that has magnitude and direction is known as a **vector**. We label the vector E, for *error*, because it represents the amount and direction by which the reading missed the true point. Usually we don't know E, but here we can calculate it. The following discussion indicates how.

In general, when we have used GPS, we have used the reported coordinates, O, as an approximation of T. The vector E was the (unknown) amount by which we missed determining the position T. As an equation we could write

$$T = O - E$$

where we record O and we disregard E to find an approximation of T. That is, the true coordinates are the observed coordinates minus the error. At best, we could estimate

---

* This discussion provides information about the result of the differential correction process. Actual techniques are involved and varied.

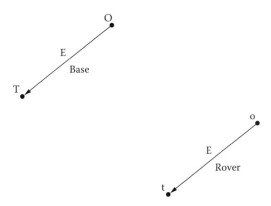

FIGURE 4.5 Known error vector applied to point observed by rover.

the magnitude—but not the direction—of *E*. (It is important to realize that none of these quantities are **scalars** (simple numbers like 23.5) but are three-dimensional entities, so the "–" sign indicates vector subtraction. The concept we are attempting to communicate survives this complexity.)

But if we know *T* exactly, and of course we have the measured value *O*, then we can rewrite the above equation to find *E*:

$$E = O - T$$

What good is being able to calculate *E*? It allows us to correct the readings of *other* GPS receivers in the area that are collecting fixes at unknown points.

We demonstrated above, with Figures 4.1, 4.2, and 4.3, that if two GPS receiver antennas are close, and use the same satellites, they will perceive almost the same errors. That is, for any given point at any given moment, *E* will be almost exactly the same for both receivers. Thus, for any nearby point reported by a GPS receiver as *o*, its true value *t* can be closely approximated simply by applying the following equation:*

$$t = o - E$$

Because both *o* and *E* are known, the error is effectively subtracted out, resulting in a nearly correct value for *t*, as shown by Figure 4.5.

This technique provides an opportunity for canceling out most of the error in a GPS position found by an antenna that is close to another antenna which is over a known point. As I mentioned before, "close" is about 500 kilometers or 300 miles. The formula for the amount of error you might expect with differentially corrected data is dependent on the distance between the base station antenna and the rover antenna. A rule of thumb is that the fix will be in error by one additional centimeter for each three kilometers between the two antennas. This relationship is approximately linear: three hundred kilometers would produce error of about a meter.

---

* Capital *O* and *T* are used to indicate base station variables; lowercase *o* and *t* indicate a rover.

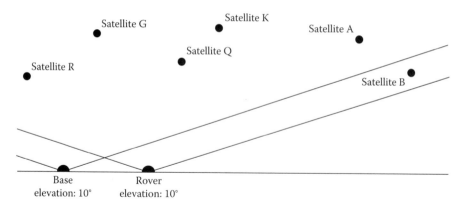

FIGURE 4.6 Base station misses a satellite (B) that the rover sees.

MAKING DIFFERENTIAL CORRECTION WORK

From a practical point of view, a number of conditions have to be satisfied for the process to work. The base and rover have to be taking data at the same time, and the base has to be taking data frequently. If the base station is to serve the rover, wherever it may be (within the 500 km limit), regardless of when data are taken, the base station must take data from all satellites the rover might see. This can cause a problem: if the base and rover are widely separated the rover might see a satellite that the base cannot view. For example, if the base and rover use the same elevation mask (say 10°), the rover might see "Satellite B" in Figure 4.6, while the base would not.

Usually base stations are set up with an elevation mask value of 10°. A good rule of thumb is to increase the elevation mask for the rover by one degree for every 100 kilometers (60 miles) it is from the base station. So a setting of 15 degrees works well, for the rover in general if it is within 500 km of the base station. Fifteen degrees is also good for avoiding difficulties due to terrain and that elevation angle reduces errors that tend to occur when signals come from satellites low on the horizon, since those signals must pass through more atmosphere. In any event the rover elevation mask must be set high enough so that there is no possibility the rover will record data from a satellite that the base is not recording, as illustrated by Figure 4.7.

Correcting errors by the differential correction method implies that the base station—the receiver at the known point—can communicate with the roving receiver(s). In practice this happens either:

- after the readings from the receivers are loaded into a computer, which we call post-mission differential correction, or
- at the time the readings are being taken, called Real-Time Differential GPS (RDGPS).

In RDGPS, a radio link is set up between the base station and the rover. As soon as the calculation for a given point is completed by the receivers, a rover uses the correction signal broadcast by the base station to adjust its opinion of where the point is.

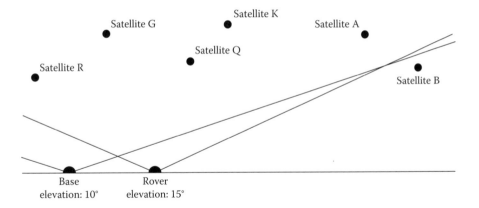

FIGURE 4.7   Neither base station nor rover sees Satellite B.

In **post-processing** GPS (more correctly called **post-mission-processing** GPS), the data from both the base and the rover are brought together later in a computer, and the appropriate correction is applied to each fix created by the rover. In the projects that follow, you will use post-mission processing to correct GPS files. For mapping and GIS activities this approach yields better results.

PROOF OF THE PUDDING

The above discussion became pretty theoretical. Is differential correction worth it? You be the judge. The files we used to open this Overview are shown again below. But there is something additional: a "smudge" in the lower left of Figure 4.8. Here we added all the fixes that were displayed before, but here each fix was differentially corrected. In other words, all the fixes with their obvious errors are now within the area of one small circle—that presumably includes the true location of the antennae. The ticks again are 10 meters apart, so you can get an idea of the amount of error reduction and the accuracy you can expect from differentially corrected data.*

REAL-TIME, DIFFERENTIAL GPS POSITION FINDING

The focus of this text is using GPS to produce input for GIS. Above we discussed what is called post-processing correction in some detail. While you can get more accurate fixes from such a system, it requires a number of additional, time-consuming steps, after data collection. Once you know the source of the differential corrections file, you have to get it, load it onto your computer, and execute software to produce the final corrected file. Wouldn't it be nice if the GPS unit simply gave accurate positions at the time you took the data? Further, there are some applications in which there simply isn't time to post-process GPS fixes—bringing a ship into a narrow harbor, for one example.

---

* Probably the term should be *differentially adjusted data*, because the resulting data still contains errors. Some of the inaccuracy has been taken out, but the results are by no means exactly correct.

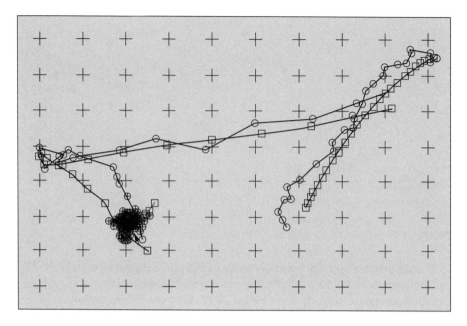

FIGURE 4.8   Eight files: four as collected, then as corrected.

I mentioned before, there is a way to provide instantaneous fixes which are almost as accurate as those created by post-processing differential correction: Real-Time Differential GPS.

GETTING CORRECTIONS FOR GPS MEASUREMENTS—RIGHT NOW!

Real-Time Differential GPS (RDGPS) may be approached in several ways:

1. The user's GPS receiver can receive correction signals from a communications satellite parked over the equator. These signals come from data taken by base stations located in disparate parts of the United States and other places in the world. The data are analyzed, packaged, and sent to geostationary communication satellites for rebroadcast to Earth. Satellite Based Augmentation Systems (SBAS) are the most important and ubiquitous of these.

2. The user can set up a base station over a known point and arrange for transmission of a radio signal from the base station to roving receivers, using a separate transmitter from the base station and a separate receiver connected to the GPS unit.

3. The user's GPS setup can receive correction signals broadcast from an antenna in the area operated by a corporation or a government agency. Such installations operate a base station continuously and broadcast the correction data, sometimes as part of another signal such as a standard FM broadcast. The National Geodetic Survey (NGS), an office of NOAA's National Ocean

Service, coordinates a network of Continuously Operating Reference Stations (CORS). Each CORS site provides GPS (and GLONASS) carrier phase and code range measurements that support three-dimensional positioning activities throughout the United States and its territories. Companies such as Accupoint, Inc., and Differential Corrections, Inc., sell a correction service.

With all three of these methods, many of the requirements for base stations remain the same. They must take data from all satellites that the rover might use. They must take data frequently (every few seconds). They should have a separate channel for each satellite so as to track it continuously.

With all the methods of doing real-time differential correction, you might wonder how these pieces of equipment manage to talk to each other. While the frequencies on which broadcasts take place may vary, the content of real-time differential correction data is standardized. The current standard is RTCM SC-104 (version 2), promulgated by the Radio Technical Commission for Maritime Services, in Washington, D.C.

Now let's look in more detail at the three steps for real-time differential correction.

SBAS Availability

Most important for using this text is the first method referenced above, because the receivers we are using are capable of using the signals for database building and because of the development of Satellite Based Augmentation Systems (SBAS) by several government agencies worldwide. These are systems that cover large regions, such as the United States or Europe. There are several of these that have come online since 2003, mainly to increase flying safety.*

In the United States is the Wide Area Augmentation System (WAAS), operated by the Federal Aviation Administration (FAA). Europe has EGNOS (The European Geostationary Navigation Overlay Service), operated by the European Space Agency. In Japan there is the Multi-functional Satellite Augmentation System (MSAS) system.

Additionally, there are commercial systems run by such corporations as Furgo (Omnistar, primarily for oil exploration) and John Deere (Starfire, for agricultural applications).

The governmental SBASs provide instantaneous correction of GPS signals. Further, they broadcast on the L1 GPS navigation frequency of 1575.42 MHz, so that no second receiver is required to get the correction signals. Both the Trimble and Magellan receivers you may have used in Chapters 1 and 2 have the capability of receiving SBAS (WAAS, EGNOS, etc.) broadcasts, and therefore the capability to provide differentially corrected data as a primary output. To do so, however, the antenna needs to be able to see a number of GPS satellites, but also one SBAS geostationary satellite.

---

* As you are quite aware by now, the vertical component of a GPS position is much less accurate than the horizontal one. And one might argue that the vertical component is, in many instances, the more critical and dramatic one. Think about being on the ground 10 feet away from a ladder compared with being 10 feet UP on the ladder and you will get the point. Also know that, in an airplane, a miss of the proper altitude at the point of landing of more than a meter or two is a sort of controlled crash. A miss by more than that is a real crash.

Recall that the geostationary satellites are all in the same orbit, remaining motionless with respect to any position on the Earth, and hovering directly over the equator in what is sometimes called the Clarke belt.* It is more difficult, especially from the ground, to acquire the signals of a SBAS satellite. Aircraft have an easier time of it, of course, having a clear view of the sky above them, and usually a direct view of the Clarke belt (35,786 km—22,236 mi) above the Earth.

COMMERCIAL SATELLITE SYSTEMS FOR REAL-TIME DIFFERENTIAL CORRECTION

A commercial solution to the real-time differential problem is also a system in which the broadcast of the correction data comes from a communications satellite. This means that, for a large portion of the Earth's surface, the user's receiver is never out of range in the area covered by the satellite, though local topography may block the signal. The satellite of one system that currently employs this method (OMNISTAR) might be as low as 30 degrees above the horizon in northern parts of the United States. The satellite broadcasts a straight-line signal of about 4000 megahertz. The area covered is all of the United States (except parts of Alaska) and parts of Canada and Mexico. Another system that provides a similar service is Racal Landstar. Both systems provide coverage over large parts of the Earth.

OMNISTAR has about a dozen base stations located on the periphery of the United States. These stations transmit data regarding the errors in the GPS signals in their areas to a central network control center. These data are then analyzed and repackaged for transmission to the communications satellite. The system uses this approach with a satellite located on a meridian that passes through Lake Michigan.

The broadcast from the satellite is such that the data from the several stations can be tailored to the user's position. How does the OMNISTAR system know where the user is? The user's GPS unit tells its coordinates to the attached OMNISTAR radio, which then decodes the signal from the communications satellite to provide the proper corrections for the local area. Since real-time differential correction information is useless unless it arrives at the GPS receiver within moments of the time a fix is taken by the receiver, you can see that a lot has to happen in an extremely short amount of time. (A GPS signal code leaves a GPS satellite; it is received by your receiver and also by the OMNISTAR base stations; the base stations transmit the signal to a central location where it is processed and sent to the communications satellite, where it is resent to the radio next to your GPS receiver; that radio determines the correction your GPS receiver needs and supplies it. Whew!)

This system can provide measurements such that 95% of the fixes taken lie within half a meter of the true point and the mean value of a number of points is accurate to within a centimeter (horizontally), using a top-of-the-line survey grade receiver. With a GPS receiver, you can expect 95% of the fixes to lie within 9 meters, with means within 2 meters of the true point.

---

* Arthur C. Clarke, the science and science fiction writer.

A User-Operated Real-Time Base Station

The user may operate her or his own base station. Those who need really precise coordinates relative to some nearby known point (e.g., land surveyors) use this method frequently. The complications go up with differential correction (real-time or otherwise), of course. Here the user has to have at least two GPS receivers, plus a transmitter associated with one of them (the base station) and receivers for each rover. In addition to the roving GPS receiver and all of its settings and conditions for good position finding, the user has to also be concerned about the same factors for the base station, plus making the radio link between the stations function at some distance. Additional complications relate to mounting the base station antenna at a precisely known point.

Centrally Located Real-Time Base Stations

Since GPS signal errors tend to be quite similar over wide geographic areas, there are obvious advantages to having a single base station serve for any roving stations in that area. Put another way, one can think of very few, if any, reasons for each of 28 GPS users, who are in reasonably close proximity to each other, to collect and rebroadcast identical correction data. This obvious fact, plus the entrepreneurial nature of American society and governmental interest in promoting spatial accuracy, has produced GPS differential correction services, wherein a user contracts for equipment and the right to receive corrections to the raw GPS signal.

STEP-BY-STEP

Project 4A—Differential Correction with SBAS (WAAS, EGNOS, or MSAS)

This is a project that requires you to (1) be outside, and (2) have your GPS receiver with you. If that doesn't work out at the moment, skip this project and come back to it later. The projects that come after this one do not require outdoor activity.

The goal of this project is to demonstrate that your GPS receiver can (or maybe cannot) differentially correct positions using a Satellite Based Augmentation System (SBAS), e.g., WAAS in and around the United States, EGNOS in and around Europe, and MSAS in the area of Japan.

One point to make up front: post-processing differential GPS is usually somewhat more accurate than real-time DGPS. However, sometimes you can collect data with an SBAS correction, and then further correct that data with post processing.

What follows are the procedures to see if you can in fact receive and process SBAS signals. Find a location that is clear, particularly (if you are in the northern hemisphere) from the southeast, through the south, to the southwest. In the southern hemisphere you would want the area to the north to be clear. The idea, wherever you are, is to have a good shot at the area above the equator.

Juno ST with TerraSync: Turn on the Juno. Make sure that GPS is active. Go to Setup > Real-time Settings. You will find a drop-down menu. If there is only one choice it will be to Use Uncorrected GPS. Click the down arrow and choose Integrated SBAS. Choice 2 will automatically become Use Uncorrected GPS. (If you now click that you can get a second choice of Wait for Real-time, but let's not use that.) Now you will see a flashing antenna next to the battery symbol. The flashing means that the receiver is not yet receiving SBAS signals. Now, in Setup, change the logging settings Style to Time with an interval of 5 seconds. Go to Data and Create a Rover file. Wait. Sometime soon, one hopes, you will have enough GPS satellites to begin collecting data. Also soon the little antenna will stop flashing and the data points you are collecting will be differentially corrected by the SBAS. Walk around a bit. After a couple of minutes close the file.

Return to a computer that is running Pathfinder Office and upload the GPS file. Display the Map (un-joined points) and Position Properties. As you click on the data points you will find some that say 3D Uncorrected and others that say 3D Real-time SBAS.

ArcPad: Turn on the Magellan MobileMapper (or whatever receiver you have ArcPad running on). Start ArcPad. Activate GPS. Start a new QuickProject. Tap the pencil and turn off Lines. Tap the pencil and turn off Polygons. Set the capture preferences so that three points will be averaged. Wait until the GPS Position window indicates "3D." Click the "satellite to single point" icon. Name the point S1. Walk a bit and collect another point—call it S2. Collect S3. At some point, you hope, the GPS Position window will indicate "DGPS 3D." When it does collect a point, calling it S101. Collect another, calling it S102, and so on. Exit QuickProject by starting a new QuickProject.

Return to a computer that is running ArcGIS. Upload the QuickProject to a folder of your choosing. Start ArcMap. Add the Points in the QuickProject folder. Use the Identify tool on the points. S1, S2, and so on are uncorrected 3D points. S101, S102, and so on have been adjusted with SBAS.

See the documentation for your GPS receiver to determine if it receives SBAS signals and, if so, how to activate the capability. Record the information in your My_GPS folder.

PROJECT 4B—BACK PORCH

Look at the Effects of Differential Correction

Pathfinder Office: Before you post-process any data yourself, you will see a demonstration of the effects of differential correction. You will consider five files:

1. An uncorrected SSF file taken with the antenna stationary. It is named R100513B.SSF and it was collected as a Not In Feature type file. It is located in __:\GPS2GIS\BP_2008, as are the other files in this project.
2. Each fix of the file above was modified by differential correction; the corrected file is called R100513B.COR.
3. R100513B.COR was then converted, using the Grouping Utility, to R100513B_COR_LINE, a Line_generic file, so it can be distinguished when you display it.
4. R100513B_COR_AVG: A Point_generic file consisting of a single point, which is the average of the corrected fixes—that is, the average of the fixes in R100513B.COR.
5. R100513B_NOTCOR_AVG—average of the uncorrected fixes.

{__1} *Start the Pathfinder Office software. If you are asked about a Project, select Default. (We won't be using the advantages of a Project in this exercise. "Project" is a convenience for pointing to a set of files and folders, but it is not necessary for the operation of Pathfinder Office. Since we will be jumping around from folder to folder, we will put aside the Project approach for the moment.) Close any open windows.*

{__2} *Click the "Open" icon (a yellow file folder on the Standard Toolbar) and navigate so you can see the files in __:\GPS2GIS\BP_2008. Set "Files of Type" so you can see "**All files *.***". (I did away with the SSF and COR file extensions when I converted some files.) There you will find R100513B.SSF. Single click on the file icon. Note that the file consists of 101 observations taken within a single-hour period; the antenna was stationary. On what date was the data set collected? _____ You will also find R100513B_COR_LINE, also consisting of 101 fixes—actually the same ones, except that each fix has been moved by a correction vector such as we described in the previous overview. Finally, R100513B_COR_AVG. It is a single point that is the average of R100513B.COR. Open all three files. Even if the Map window is open, don't bother looking at them yet.*

{__3} *Set the Coordinate System to UTM, Zone 16 in the Northern Hemisphere. Set Coordinate Units to Meters. Set the Datum to WGS 1984. Set "Altitude*

Measured From" to MSL (EGM96 Global) in meters. Set Distance units to Meters. The time zone is Eastern Daylight.

{__4} Under View ~ Layers ~ Features set the following:

(Make sure the "Join" box is checked so you get lines and not just points, which are less easily visible.)

Not In Feature: a thin gray line (for R100513B.SSF)
Line_generic: a thin green line (for R100513B_COR_LINE)
Point_generic: a red square (for R100513B_COR_AVG)

{__5} Open both the Map and Time Line windows. Open the Position Properties window (using either the icon or the Data menu). Examine the trace made by the original data file, shown by gray lines, and the corrected data file, shown by the green lines. Notice the (approximate) congruence of the two sets of lines. In the southeast corner you will notice two little triangles made by the paths connecting the points. Click on each point of the north apex of each triangle; you can tell from position properties that it is point number 16 of its respective file. How far apart are these two points? _____ meters. They are pretty close together, given that one is 3D Uncorrected and the other is Postprocessed Code. Here is a clue as to why the raw data is so much in agreement with the corrected data: Click on the DOPS tab of the Position Properties window. How many satellites went into finding point 16? _____ What was the PDOP of the raw data? _____ That is to say, conditions for taking data at that moment were very good, allowing over-determined position finding, so the raw data approximated the corrected data.

{__6} Press the Summary tab in the Position Properties window. Zoom in on the northernmost green line. Verify that the fixes at the ends of this line are 88 and 89. Barely south of this line is a fix from R100513B.SSF (it is number 78). Click on that fix. Note that it is described not as 3D Uncorrected but rather as 3D Real-time SBAS. So some of the points in R100513B.SSF were adjusted by a Wide Area Augmentation System (WAAS) satellite as they were placed in the receiver's memory, and some were not. The differential correction program takes this into account when post-processing data. Zoom to Extents.

{__7} Using the Select arrow pointer, click on the single point of R100513B_ COR_AVG in the Map window. A box should appear around the point in both the map and time line windows. If you have set all the parameters correctly, the Position Properties window should reveal the location of this average point:

4,207,342.015 meters (north of the equator) and,

720,648.399 meters (which is 220,648.399 meters east of the 87th (west) meridian, which is the central meridian of UTM Zone 16)[*]

The altitude: 295.714 meters.

{__8} *Now open these three files again and include R100513B_NOTCOR_ AVG, which is the average of these uncorrected fixes. For these fixes fill out the blanks below:*

Northing _____ meters (north of the equator),
Easting _____ meters (east of the 87th (west) meridian,
Altitude _____ meters.

Use measure to determine how far apart are the averages of the two files are horizontally? _____ Looking at position properties, what is the vertical difference? _____

If you have ArcMap available continue with the exercise immediately below.

---

[*] If this confuses you, you need to do a bit of study on the UTM coordinate system.

ArcMap: Before you post process any data yourself, you will see a demonstration of the effects of differential correction. You will consider six files:

1. A shapefile taken with the antenna stationary. It is named BP_all.shp and it resides in IGPSwArcGIS\BP_2008, as do all the files in this project.
2. Each fix of the file above was modified by differential correction with signals from a base station; the corrected file is called BP_all_base_ corrected.shp.
3. A shapefile consisting of a single point that is the average (mean) of the points taken in (1) above, called BP_average_uncorrected.
4. A shapefile consisting of a single point that is the average of the points taken in (2) above, called BP_all_average_corrected.
5. The points in BP_all that were not corrected by SBAS (WAAS), called BP_not_SBAS_corrected.
6. The points in BP_all that were corrected by SBAS, called BP_SBAS_corrected.

{__9} *Start ArcMap. Add as data*

IGPSwArcGIS\BP_2008\BP_all.shp.

Right click on the symbol in the Table of Contents (T/C) and make the symbol black. With the Units set to meters, measure the distance separating the two fixes farthest apart. _____

Open the attribute table of BP_all.shp. Note that the file consists of 101 observations taken within a single-hour period. Run Statistics on the GPS_Height. What is the average height? _____ What is the Standard Deviation? _____

{__10} Turn off BP_all.shp and dismiss its table. Add as data

IGPSwArcGIS\BP_2008\BP_all_base_corrected.shp.

Make the symbol green. With the Units set to meters, measure the distance separating the two fixes farthest apart. _____

Open the attribute table of BP_all.shp. The file consists of the same 101 observations, but adjusted by a nearby base station. Run Statistics on the GPS_Height. What is the average height? _____ What is the Standard Deviation? _____ Dismiss the table.

{__11} Add as data the two files BP_average_uncorrected.shp and BP_average_corrected.shp. These files represent the three-dimensional averages of the files that contain the 101 fixes. Left click on the 'uncorrected' symbol in the T/C and make the symbol Octagon 1 (black). Make the 'corrected' symbol Circle 2 (green).

Measure the distance between the two. _____ The 'corrected' symbol should be closer to the true location.*

The remaining two files are instructive because one consists of fixes taken when SBAS was not available. There are also fixes which were corrected by SBAS.

{__12} Turn off all four files. Add the files BP_not_SBAS_corrected.shp and BP_SBAS_corrected.shp. Use different symbol colors so you can easily distinguish one from another. You can tell that the SBAS corrected fixes are more tightly clustered.

---

* We do not actually know the true location; we are simply inferring that it lies close to the average of the corrected points.

An Important Notice

You can do Projects 4C and 4D in this chapter only if you have Pathfinder Office available. In general, if you are using data collected by ArcPad you should be sure that SBAS is active when you collect fixes and that you average, say 30 to 80, fixes to find a given point.

PROJECT 4C—THE MCVEY MONUMENT

Correct Some Supplied Point Data

In the steps below you are supplied with a data file of 60 fixes taken with the antenna placed over the carefully surveyed McVey National Geodetic Survey (NGS) marker located on the University of Kentucky campus.* You are also supplied with base files taken by Trimble community base stations. Base station files usually consist of an hour's worth of position data. A new file is started every hour, on the hour, 24 hours a day, seven days a week.

The format of a base station file is XYMMDDHH.SSF, where

- "X" is an arbitrary, permanent "prefix" letter chosen by the base station managers to identify files from that base station
- "Y" is the last digit of the year in which the base station file was collected
- "MM" is the number of month in which the base station file was collected
- "DD" is the number of the day in which the base station file was collected
- "HH" is the hour, **in UTC time**, of the day in which the base station file was collected

In general you can obtain a base station file over a computer communication network such as the Internet or its alter ego, the World Wide Web. Another way is to download the file to your computer through anonymous FTP (File Transfer Protocol). Or, you could request a base station file be sent to you through the mail on a CD-ROM if the base station does not have a Web site or anonymous FTP—a situation which occurs less and less frequently. Base station files are usually large (compared to rover files)—from 150 Kb to 350 Kb. Sometimes, to reduce their size, and hence the time it takes to transmit them electronically, they are compressed into "ZIP" files or self-extracting "EXE" files. Base stations usually record a data fix every few seconds from all the satellites in view, using an elevation mask of 10°.

{__1} *Start a new Project named McVey_NGS_yis, with the primary folder as* ___*:\GPS2GIS\McVey_NGS. The folder structure has already been set up for you with Base, Backup, and Export folders.*

{__2} *Close all files and windows (except Pathfinder Office, which should occupy the full screen). Bring up the Differential Correction window, either by selecting it in the Utilities menu or by clicking on the Differential Corrections icon (it looks like a target) on the Utility toolbar.*

Because you have a Project selected, navigating to the proper files should be easy. Basically, to fill in the blanks of the DIFFERENTIAL CORRECTIONS window, you have to select

- the rover file(s) that you want corrected,
- the base file(s) needed to do the corrections, and

---

* The National Geodetic Survey (NGS) is part of the National Oceanic and Atmospheric Administration (NOAA). It is the United States' oldest scientific agency; Thomas Jefferson established it in 1807 as the Survey of the Coast.

- where the corrected file(s) are to be placed.
- some parameters

{__3} *Under Rover Files click Browse. Navigate, if necessary, to the file* ___:\GPS2GIS\McVey_NGS\M032522A.SSF. *(If it doesn't appear, make sure Files of Type include those files with SSF extensions.) Click on the file. Click on "Open" to okay the choice. Back in the Differential Corrections window, the choice should appear; the folder and file name should both be presented.*

{__4} *Under Base Files click Browse. The contents of the "base" folder, located within the McVey_NGS folder, should appear, containing a set of filenames of base station files. Among others you will see four files—two from each of two nearby base stations. They are:*

A9032521.SSF
A9032522.SSF
L9032521.SSF
L9032522.SSF

The files that begin with "A" are from the UK Department of Agriculture. Those starting with "L" are from the Bluegrass ADD—a regional area development district. Two of these files might be appropriate for correcting M032522A.SSF. Which are they?

_____ _____

{__5} *Cancel the "Select Base Files" window. (We have an alternative way to get them.)*

{__6} *Click Local Search to initiate a search of the project folder Base folder within the McVey_NGS folder. A "Local Search for Base Files" window will appear, looking like Figure 4.9. It should show* ___:\GPS2GIS\McVey_NGS\Base *in the Folder text window.*

{__7} *After making sure that the path to the folder is correct, place an "A" in the Preferred Base Station Prefix field. Then press Search.*

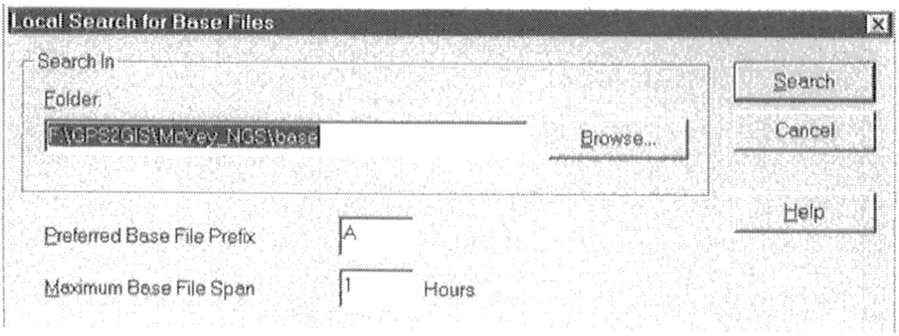

FIGURE 4.9   Finding differential correction files in the Base folder.

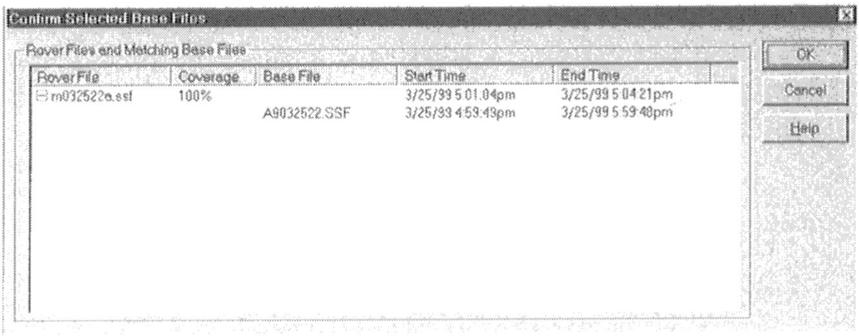

FIGURE 4.10 Checking compatibility of the rover and the base station files.

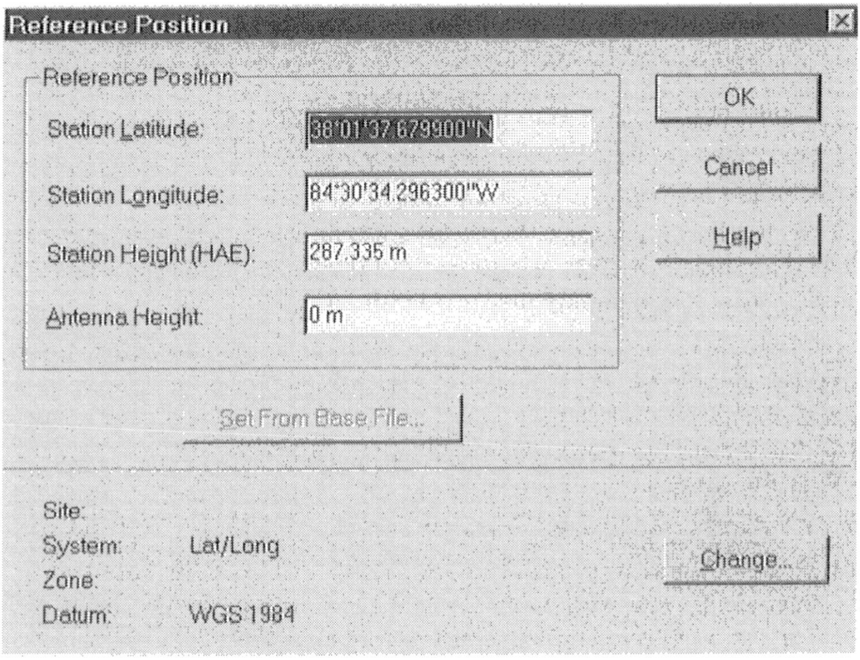

FIGURE 4.11 Reference position of the base station.

{__8} *A window asking you to Confirm Selected Base Files will appear—please see Figure 4.10. Study this window. Assure yourself that the time covered by the base station file indeed includes the time period over which the rover file was taken. Then okay the window.*

{__9} *A window showing the Reference Position of the base station will appear. See Figure 4.11. You could put in new coordinates, but you have no reason to dispute that what is shown is indeed the location of the base station antenna. You could change the coordinates, but don't. Okay the window.*

{__10} *Check to see if the Output Folder specifies ___:\GPS2GIS\McVey_NGS and that the file extension is "COR." The name of the corrected file will be M032522A. COR. Since you will be correcting "code" data (as differentiated from "carrier" data) be sure that either "Smart Code and Carrier Phase Processing" or "Code Processing Only" is checked under Processing. Look over the Differential Corrections window; okay it if it is correct; fix it if it's not.*

{__11} *You may get a warning that a file will be overwritten. You want to continue anyway.*

{__12} *Watch the reports of the progress of the correction process. When it is finished 60 positions should have been adjusted; ask for "More Details." Examine the resulting report, then dismiss it. Close the Differential Corrections Completed window.*

Map the Data

{__13} *Open the files M032522A.SSF and M032522A.COR. Map them. Unfortunately you can't tell much because they are both Not in Feature files. Do you recall how to use the Grouping utility to convert a file from a Not In Feature file into a Generic_ line feature file? Do you recall also how to get a Generic_point feature file, that averages the 60 fixes to get the estimate of the single point? If not, use the next few steps to guide you.*

{__14} *Start the Grouping utility. For the input file, "Browse" and choose M032522A. COR. Call the output file MV_LINE.COR. Use "One group per input file" and create the feature type* **"Lines."** *Okay.*

{__15} *Start the Grouping utility again. For the input file choose M032522A.COR. Call the output file MV_POINT.COR. Use "One group per input file" and create the feature type* **"Points."** *Okay.*

{__16} *Open M032522A.SSF, MV_LINE.COR, and MV_POINT.COR. If necessary, fix up the display parameters with View ~ Layers ~ Features so that you can tell the three files apart. Again, note the improvement that differential correction made. Enlarge the Map window somewhat and Zoom up on the corrected fixes.*

{__17} *Open the Time Line window. With the arrow pointer, select MV_POINT. COR so that there is a box around it. Under Options, make the coordinate system Latitude and Longitude and the Datum NAD 1983 (Conus). The altitude units should be U.S. Survey Feet. Make the Style of Display DD*MM'SS.ss"—that is, degrees, minutes, seconds, and decimal fractions of a second.*

{__18} *Open the Position Properties window. Write below the position of the point and compare it with the published coordinates based on the NAD 1983 datum:*

Latitude: _____	Longitude: _____	Altitude: _____
38* 02' 20.45143 N	*84* 30' 18.90173"W*	*1011.965 USft*

Latitude: 38* 02' 20.57581"   *Longitude: −84* 30' 18.67673"*   *Altitude: 974.5 feet*

Differences: _____	_____	_____
0.12438	*0.225*	*37.465*

{__19} *A second of latitude is about 30.75 meters. A second of longitude (at this latitude of 38°) is about 24.36 meters (i.e., 30.75 times the cosine of 38°). What are the differences for northing, easting, and altitude in terms of meters? A meter is equivalent to 39.37 inches, based on U.S. survey feet.*

Differences (meters): _____ _____ _____

The approximate answers are 3.8, 5.5, and 11.4 meters.

{__20} *To verify that I have given you the correct coordinates for the McVey monument, look it up on the Internet. The McVey monument's designation is HZ0339. The Web site is http://www.ngs.noaa.gov. Locate the place on the Web page that allows you to "Find a point."*

To review: You just learned how to differentially correct an SSF file. What you did is called post-mission processing, or post-processing, because considerable time elapsed from the time the file was recorded in the GPS receiver and the time it was corrected by you. As mentioned previously, another way to improve GPS data is called "real-time differential correction," whereby the base station transmits a radio signal that is received by the rover(s), allowing them to correct their data almost instantly. This allows the roving receivers to display the more nearly correct position fixes immediately. One might use such a capability to bring a ship into a harbor. The objective is to miss the rocks at that moment; post-processing would hardly be appropriate. We will spend more time discussing the real-time differential correction method in Chapter 7. But for using GPS data in a GIS, **post-mission differential correction is the way to go.**

Project 4D—McVey Waypoint

Waypoints—Using One to Check Your Work
The McVey monument is a good candidate for a waypoint since it is a known reference point. We will do this in order to see another way of calculating the distance from a corrected average point to a known point.

{__1} *Still in the ___:\GPS2GIS\McVey_NGS folder you will find a file named V062318C.SSF generated at the McVey monument in June of 2000. (Click on the folder icon on the Project toolbar.) This is a file taken after SA was discontinued in May of 2000. Also, if you navigate to the "Base" folder, you will find L0062318. SSF. Close the folder.*

{__2} *Correct V062318C.SSF with L0062318.SSF.*

{__3} *Make V062318C.SSF produce a Generic_point file named V062318C_PNT. SSF.**

---

* Shortcut hint: If in the Output File window you click on a file name, that name will replace "grouped. xxx," saving you typing.

{__4} *Make V062318C.COR produce a Generic_point file named V062318C_PNT. COR.*

{__5} *Make V062318C.COR produce a Generic_line file named V062318C_LINE.COR.*

{__6} *Display the two point and one line file. With the pointer cursor select each of the point file symbols. Note which name goes with which. We could again compare the results of this operation with the published coordinates by the pencil and paper method of Project 4B. But let's look at a more elegant and efficient way. We will establish a waypoint to represent the McVey marker and display the waypoint at the same time as the three files above.*

Make a Waypoint in Pathfinder Office

{__7} *Make certain the Options are set right for defining the waypoint: Latitude and Longitude, NAD 1983 (Conus), Altitude Measured from MSL, Altitude Units and Distance Units: U.S. Survey Feet, Style: DD*MM'SS.SS.*

{__8} *Under File click Waypoints, then New. In the New Waypoint File window, navigate to the ___:\GPS2GIS\McVey_NGS folder. This window will allow you to make a file of as many waypoints as you want. The set of waypoints will appear and disappear from the screen together. In our case we will make only one file containing one waypoint. The window will suggest that you call the new waypoint file something like w<date>a.wpt but change it to McVey_yis.wpt.* *

{__9} *Okay the New Waypoint File window and another window will appear named Waypoint Properties. Punch Create and a third window, Create Waypoint, will be displayed.*

{__10} *Make the waypoint name McVey. Type in the latitude, longitude, and altitude as shown by the alphanumeric strings below:*

*38*02'20.57581"*
*−84*30'18.67673"*
974.5

The Create Waypoint window should now look like Figure 4.12.

{__11} *Press Save, and then Close in the Create Waypoint window. The Waypoint Properties window will reappear. In your file McVey.wpt you have only one waypoint: McVey. Check that the coordinates of McVey are correct. If not, Edit them. When done, close the Waypoints Properties window.*

---

* Where *yis* indicates your initials—you know the drill.

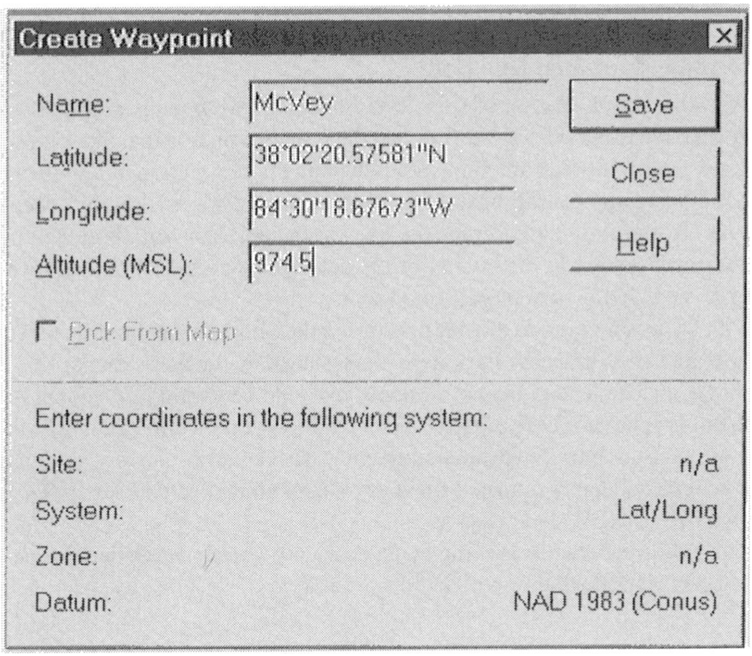

FIGURE 4.12   Creating the McVey waypoint.

{__12} *If it is not displayed, open the waypoint file: File ~ Waypoints ~ Open ~ McVey_yis.*

{__13} *Select a blue triangle to represent the waypoint under View ~ Layers ~ Waypoints ~ Symbol. Make the Map window active. The newly created waypoint may appear in the Map window. If it doesn't, click on the Zoom Extents icon (magnifying glass with the equal sign (=) in it).*

{__14} *Set the distance units to meters.*

{__15} *Measure the horizontal distance from the COR point to the waypoint. _____. To the SSF point. _____*

{__16} *With Data ~ Measure, determine the approximate north-south distances from the data point to the waypoint. COR:_____SSF: _____*

{__17} *Do the same for the east-west distance. COR: _____. SSF: _____*

{__18} *And then for the altitude. COR: _____ SSF: _____*

PROJECT 4E—VANCOUVER

Look at Some Corrected Line Data

Next we want to look at a set of some line files that have been post-processed. The primary difference between these files and the previous ones is that these fixes represent a track from a moving antenna, rather than an approximation of a single specific point. Obviously, the overall accuracy of the corrected file will be less, because it would not be appropriate to average the fixes in the file. You will, however, notice a dramatic improvement in the quality of the data collected by a "roving" receiver—whether it is carried by automobile or on foot.

The files you are going to look at first were taken both by automobile and by hiking in the area of Whytecliff Park near Horseshoe Bay, in West Vancouver, British Columbia, Canada. A road map of Vancouver would be useful (are you an AAA or CAA member?) but not necessary. Or you could look up the area on one of the map Web sites, such as http://www.mapquest.com. The created files are designated as OLD. The data collectors returned a few years later and repeated the trip. These files are designated as NEW.

You can choose between looking at the data with Pathfinder Office (immediately below) or ArcMap (further down).

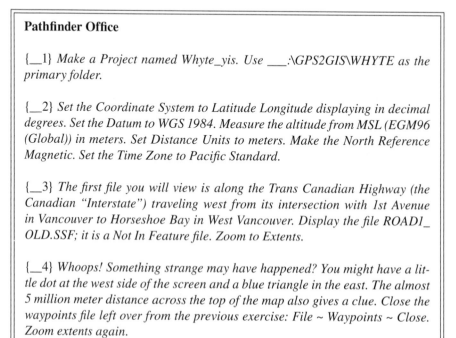

Pathfinder Office

{__1} *Make a Project named Whyte_yis. Use ___:\GPS2GIS\WHYTE as the primary folder.*

{__2} *Set the Coordinate System to Latitude Longitude displaying in decimal degrees. Set the Datum to WGS 1984. Measure the altitude from MSL (EGM96 (Global)) in meters. Set Distance Units to meters. Make the North Reference Magnetic. Set the Time Zone to Pacific Standard.*

{__3} *The first file you will view is along the Trans Canadian Highway (the Canadian "Interstate") traveling west from its intersection with 1st Avenue in Vancouver to Horseshoe Bay in West Vancouver. Display the file ROAD1_ OLD.SSF; it is a Not In Feature file. Zoom to Extents.*

{__4} *Whoops! Something strange may have happened? You might have a little dot at the west side of the screen and a blue triangle in the east. The almost 5 million meter distance across the top of the map also gives a clue. Close the waypoints file left over from the previous exercise: File ~ Waypoints ~ Close. Zoom extents again.*

{__5} *Open whatever windows you need to answer the following questions. How long was the trip in minutes? _____ In kilometers (which are standard in Canada) as the crow flies? _____ Along the road? _____. By using a vertical line with the measure command, estimate the difference between True North and Magnetic North in southwestern Canada. _____°.*

{__6} *Minimize the Pathfinder Office window (by clicking on the little bar icon near the upper right corner). Close any other windows you need to in order to get to the Desktop. Using the Windows Explorer or "My Computer," bring up a window containing a list of files in the ___:\GPS2GIS\WHYTE\BASE folder. Notice a file named "W4012921.EXE." The name of this file has the format of a community base station file, but the extension is EXE, not SSF.*

W4012921.EXE is an interesting type of file. An EXE file is usually a computer program—one that may be executed by the Central Processing Unit (CPU) of your computer. This particular file is partly program, but mostly compressed data. It could be called a "self-extracting GPS data file" because when it is executed it generates a community base station SSF file.

{__7} *Execute "W4012921.EXE." You do that just by pouncing on the icon next to its name. A DOS window will appear momentarily while the new SSF file is extracted.*

{__8} *Again examine the files in the folder. You will notice that "W4012921. SSF" has been added. This is a standard storage format file from the Whiterock, BC, community base station, located near the U.S. border in southwestern British Columbia, about 53 kilometers (33 miles) from Horseshoe Bay.*

{__} *Revive Pathfinder Office by clicking its button on the task bar. Open the file ROAD2_OLD.SSF. What happened here, though you can barely tell it from the GPS track, is that the car proceeded from the southeast (on Marine Drive), went mostly west, then northeast, until the road made a large "U," headed southwest and then into the parking lot of Whytecliff Park at the far left of the track image. It cruised around the lot, and returned, going northeast, along the road it had just taken. Finally it turned left and encountered a dead end, where it parked.*

{__9} *Now open the same file but add the file ROAD2_OLD.COR, which is a file that was made by differentially correcting ROAD2_OLD.SSF. The base file that was used was W4012921.SSF (which, as you may recall, you generated by executing the file W4012921.EXE). ROAD2_OLD.COR was turned into a Line_generic file so that you could distinguish it from its uncorrected cousin. Use View ~ Layers ~ Features if necessary to display these two files differently.*

Now, if you reread the description of the car's track in the previous step, the path should be more obvious. If you want to look at a map of the area you can go to http://www.mapquest.com and put in the address 6992 Hycroft Road, West Vancouver, BC, Canada. The red star shows where the car parked.

The receiver was then carried by hikers through the park, from the dead end in the northeast to the parking lot in the southwest. Two files represent this trip: HIKE1_OLD.SSF goes from the dead end, where the car stopped at the end of the ROAD2_OLD.SSF file, up to the pinnacle in the park. HIKE2_OLD.SSF goes from the pinnacle down to the Whytecliff Park parking lot in the west.

{__10} *Display HIKE1_OLD.SSF.*

The jagged track does not represent where the hikers actually walked. The spikes in the track probably come from a combination of selective availability and frequent changes in the "best PDOP" constellation, caused by terrain interference. Contributing further to the poor quality of the track was the fact that the receiver was set to record a point each time the receiver calculated that it was 20 meters from the last point, rather than every so many seconds. This has the effect of recording any spurious distant point that is generated. Recall that the receiver calculates a fix about every two-thirds of a second; whether or not that fix is recorded in the memory depends on how you have the rover parameters set. If the criteria for recording a fix is how far that fix is distant from the last fix observed, then you can see that fixes that are distant because of errors will be recorded. If the rover is set to record a fix periodically instead, the chance that an errant fix will be recorded is reduced.

{__11} *Continue to trace the hike through the park by adding HIKE2_OLD. SSF to the set of open files. Here the GPS receiver was set to record points every 10 seconds. The track looks significantly better.*

Differential correction makes a huge difference, as you will see:

{__12} *Display the corrected files HIKE1_OLD.COR and HIKE2_OLD.COR together. These tracks aren't perfect—note some jags and double-backs—but at least you get a reasonably cohesive picture of the trail. As it turns out, not all the fixes in the rover file were corrected, which probably accounts for the spurious points remaining. Finally, display these together with ROAD2_OLD.COR. The corrected GPS tracks shown here are hardly perfect, or even very nice. But you can hardly deny that differential correction made an enormous difference in the quality.*

In May 2000 the hikers repeated the trip to Horseshoe Bay. They used better equipment and, more importantly, selective availability was turned off.

{__13} *Set the time zone to Pacific Daylight Savings. Repeat the displays indicated above but use the SSF and COR files of ROAD2_NEW, HIKE1_NEW, and HIKE2_NEW. Perhaps display the SSF files in the background and the COR files in the standard way. You will have to zoom up pretty far to see the differences in the tracks. (By the way, the car took a little side trip just before the parking area—different from the 1994 data.)*

Just off the southwest beach of Whytecliff Park is a bit of rock that is a peninsula at low tide and island at high tide. The hikers climbed this and acquired the file V052223E.SSF, which was corrected later to form V052223E.COR.

{__14} *Group each of these files to form V052223E_1PT.SSF and V052223E_1PT.COR and display them. Measure how far apart these points are from each other. _____. Write the coordinates of the corrected point in decimal degrees. Latitude: _____. Longitude (don't forget the negative sign): _____. (You will need these coordinates for the next step.)*

{__15} *Using an Internet browser, bring up http://www.mapquest.com.* Display the largest possible map at the most detailed level, using the coordinates you wrote above. The red star should show up on the island. If you zoom out to a slightly smaller scale you should be able to see Hycroft Road where the car was parked and infer the location of the hiking trail through the park.*

---

* *www*, as has been noted by others, sometimes stands for "world wide wait." You may need to be patient.

ArcGIS

{__16} *Start ArcMap. Arrange to Add Data from ___:\IGPSwArcGIS\Whyte. Make the coordinate system WGS 1984 UTM Zone 10 North.*

{__17} *The first file you will view is along the Trans Canadian Highway (the Canadian "Interstate") traveling west from its intersection with 1st Avenue in Vancouver to Horseshoe Bay in West Vancouver. Add as data the file Road1_old_uncorrected.shp.*

{__18} *Open whatever windows you need to answer the following questions. How long was the trip in kilometers (which are standard in Canada) as the crow flies? _____ Along the road? _____.*

{__19} *Add as data the shapefile Road2_old_uncorrected.shp. Zoom to this layer. What happened here, though you can barely tell it from the GPS track, is that the car proceeded from the southeast (on Marine Drive), went mostly west, then northeast, until the road made a large "U," headed southwest and then into the parking lot of Whytecliff Park at the far left of the track image. It cruised around the lot, and returned, going northeast, along the road it had just taken. Finally it turned left and encountered a dead end, where it parked.*

{__20} *Now add the file Road2_old_corrected.shp, which is a file that was made by differentially correcting Road2_old_uncorrected.shp.*

Now, if you reread the description of the car's track in the previous step, the path should be more obvious. If you want to look at a map of the area you can go to http://www.mapquest.com and put in the address 6992 Hycroft Road, West Vancouver, BC, Canada. The red star shows where the car parked.

The receiver was then carried by hikers through the park, from the dead end in the northeast to the parking lot in the southwest. Two files represent this trip: Hike1_old_uncorrected.ssf goes from the dead end, where the car stopped at the end of the Road2_old_uncorrected.shp file, up to the pinnacle in the park. Hike2_old_uncorrected.ssf goes from the pinnacle down to the Whytecliff Park parking lot in the west.

{__21} *Add as data Hike1_old_uncorrected.ssf. Zoom to this layer.*

The jagged track does not represent where the hikers actually walked. The spikes in the track probably come from a combination of selective availability and frequent changes in the "best PDOP" constellation, caused by terrain interference. Contributing further to the poor quality of the track was the fact that the receiver was set to record a point each time the receiver calculated that it was 20 meters from the last point, rather than every so many seconds. This has the effect of recording any spurious distant point that is generated. The receiver calculates a fix about every two-thirds of a second; whether or not that fix is recorded in the memory depends on how you have the rover parameters set. If the criteria for recording a fix is how far that fix is distant from the last fix observed, then you can see that fixes that are distant because of errors will be recorded. If the rover is set to record a fix periodically instead, the chance that an errant fix will be recorded is reduced.

{__22} *Continue to trace the hike through the park by adding Hike2_old_uncorrected.ssf to the set of open files. Here the GPS receiver was set to record points every 10 seconds. The track looks significantly better.*

Differential correction makes a huge difference, as you will see.

{__23} *Display the corrected files Hike1_old_corrected.ssf and Hike2_old_ corrected.ssf together. These tracks aren't perfect—note some jags and double- backs—but at least you get a reasonably cohesive picture of the trail. As it turns out, not all the fixes in the rover file were corrected, which probably accounts for the spurious points remaining. Finally, display these together with Road2_ old_corrected. The corrected GPS tracks shown here are hardly perfect, or even very nice. But you can hardly deny that differential correction made an enormous difference in the quality.*

In May 2000 the hikers repeated the trip to Horseshoe Bay. They used better equipment and, more importantly, selective availability was turned off.

{__24} *Start a new map. Repeat the displays indicated above but use the uncorrrected and corrected shapefiles of Road2_new, Hike1_new, and Hike2_new. You will have to zoom in pretty far to see the differences in the tracks. (By the way, the car took a little side trip just before the parking area—different from the 1994 data.) Turn off the Road and Hike files.*

Just off the southwest beach of Whytecliff Park is a bit of rock that is a pen- insula at low tide and island at high tide. You can see a photograph by displaying The_Rock.jpg in the Whyte folder. The hikers climbed this and acquired the file V052223E_uncorrected.shp, which was corrected later to form V052223E_cor- rected.shp.

{__25} *Display V052223E_uncorrected.shp and V052223E_corrected.shp. The fixes in these files were averaged to produce V052223E_1Pt_uncorrected. shp and V052223E_1Pt_corrected.shp. Display these as well. Measure how far apart these points are from each other. _____. Write the coordi- nates of the corrected point in decimal degrees. Latitude: _____. Longitude (don't forget the negative sign): _____. (You will need these coordinates for the next step.)*

{__26} *Using an Internet browser, bring up http://www.mapquest.com.* Display the largest possible map at the most detailed level, using the coordinates you wrote above. The red star should show up on the island. If you zoom out to a slightly smaller scale you should be able to see Hycroft Road where the car was parked and infer the location of the hiking trail through the park.*

---

* *www*, as has been noted by others, sometimes stands for "world wide wait." You may need to be patient.

PROJECT 4F

Differentially Correct Some Supplied Line Data—New Circle Road Again

---

Pathfinder Office

In this exercise you will differentially correct two files generated by a car going around a portion of New Circle Road and returning to the starting point, in September of 1999 (9/99). Most of the trip took place using the four-lane divided highway. You will also look at a DOQ of part of that highway and see the GPS tracks overlaid on it. We assume that Pathfinder Office is running as you begin to follow the steps below.

{__1} *Make a Project named CircleRd_999: Under Files select Projects. When asked to select a project click New. Give the new project the name CircleRd_999_yis. The folder structure has already been prepared for you, with base, export, and backup folders. Simply specify C:\GPS2GIS\CircleRd_999 as the primary folder. When you are done creating the project, the Current Project identifier on the Project Toolbar should read CircleRd_999_yis (or as much of it as the text window lets you see). Set the coordinate system to Kentucky state plane north, NAD 83, with units of survey feet.*

{__2} *To get an idea of where the New Circle Road is with respect to the University and Lexington, display the waypoint McVey by opening the way-point file McVey_NGS_yis. Zoom to extents. (If you didn't close the Whytecliff Park files you may get a rather large distance reported for the width of the window. Fix that by closing both the normally displayed files and the background files.)*

{__3} *Use the "Open Folder" icon (the leftmost one on the Pathfinder Office* **Project** *Toolbar) to look at the folder for the project, CircleRd_999. In this folder you will find two rover files: S090519A.SSF and S090519B.SSF. These are the files you will correct. Resize the window so it doesn't take up much room.*

{__4} *Examine the files: Use the "Open" icon (the leftmost one on the* **Standard** *toolbar) to open the files S090519A.SSF and S090519B.SSF. Make "Not In Feature" data represented by the thinnest red line. Then use View ~ Map to look at both of these uncorrected SSF files (use Ctrl-click to select both). You will notice an area at the westernmost area of the trace where the two tracks run along together along an arc. If you zoom up on that area you will be able to see that the tracks merge and cross—an indication that the GPS points contained considerable error. Return the window to the full extent of the GPS tracks.*

{__5} *Using the ___:\GPS2GIS\CircleRd_999 window* that you opened previously, pounce on the icon of the Base folder of CircleRd_999. There you should find an EXE file: U9090519.EXE. When executed, this will make the base file that you will use to correct the rover files. Pounce on the EXE file icon to execute it. An SSF file of the same name should appear in the folder. (If there is already an SSF file of a given name in the folder you will be asked if you want to overwrite it. You do; type a "Y".) Once the SSF base file is present, close the folder.*

{__6} *Start the differential correction process from either the Utilities menu or the Utilities toolbar. You will correct the two rover files with the base file both at the same time. Under Rover Files click Browse. Select the rover files by clicking on one of them with the mouse cursor, then hold down the Ctrl key and click on the other one. Click Open.*

{__7} **Select the proper base file to correct the rover files:** *Under Base Files click the Local Search button. Pathfinder Office will then automatically select the folder ___:\GPS2GIS\CircleRd_999\Base. Punch Search. Examine the Confirm Selected Base Files window then okay that window. Agree to the given Reference Position of the Base Station.*

{__8} *Under Corrected Files the Output Folder should be ___:\GPS2GIS\ CircleRd_999 and the file extension should be COR. Under Processing make sure that Code Processing will be done. Press OK in the Differential Correction window. The differential correction process should proceed, leaving you with S090519A.COR and S090519B.COR in the project folder.*

{__9} *Open the project file folder (from the Project toolbar) to see that everything is there. Pounce on one of the AUR files. Make it full screen. It will tell you more than you will want to know about the differential correction process. Dismiss the window.*

You will recall from Chapter 3 that a digital orthophoto image is an aerial photograph that has been calibrated so that it may be used as a map. Here again we make use of this capability.

{__10} **Prepare the map of the GPS track so that a digital orthophoto will be properly positioned:** *The configuration you want is the U.S. State Plane 1983 coordinate system, Kentucky North Zone (1601). The datum is NAD83 (Conus). The coordinate units, altitude units, and distance units are Survey Feet. The Time Zone is Eastern Daylight. Using the Units, Time Zone, and Coordinate System choices under the Options menu, set up these parameters.*

---

* The folder may have disappeared but it can be redisplayed by clicking its button on the task bar.

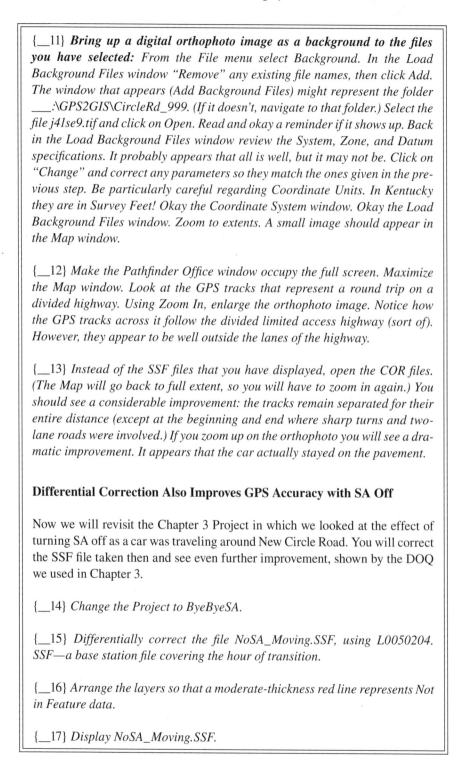

{__11} ***Bring up a digital orthophoto image as a background to the files you have selected:*** *From the File menu select Background. In the Load Background Files window "Remove" any existing file names, then click Add. The window that appears (Add Background Files) might represent the folder ___:\GPS2GIS\CircleRd_999. (If it doesn't, navigate to that folder.) Select the file j4lse9.tif and click on Open. Read and okay a reminder if it shows up. Back in the Load Background Files window review the System, Zone, and Datum specifications. It probably appears that all is well, but it may not be. Click on "Change" and correct any parameters so they match the ones given in the previous step. Be particularly careful regarding Coordinate Units. In Kentucky they are in Survey Feet! Okay the Coordinate System window. Okay the Load Background Files window. Zoom to extents. A small image should appear in the Map window.*

{__12} *Make the Pathfinder Office window occupy the full screen. Maximize the Map window. Look at the GPS tracks that represent a round trip on a divided highway. Using Zoom In, enlarge the orthophoto image. Notice how the GPS tracks across it follow the divided limited access highway (sort of). However, they appear to be well outside the lanes of the highway.*

{__13} *Instead of the SSF files that you have displayed, open the COR files. (The Map will go back to full extent, so you will have to zoom in again.) You should see a considerable improvement: the tracks remain separated for their entire distance (except at the beginning and end where sharp turns and two-lane roads were involved.) If you zoom up on the orthophoto you will see a dramatic improvement. It appears that the car actually stayed on the pavement.*

Differential Correction Also Improves GPS Accuracy with SA Off

Now we will revisit the Chapter 3 Project in which we looked at the effect of turning SA off as a car was traveling around New Circle Road. You will correct the SSF file taken then and see even further improvement, shown by the DOQ we used in Chapter 3.

{__14} *Change the Project to ByeByeSA.*

{__15} *Differentially correct the file NoSA_Moving.SSF, using L0050204. SSF—a base station file covering the hour of transition.*

{__16} *Arrange the layers so that a moderate-thickness red line represents Not in Feature data.*

{__17} *Display NoSA_Moving.SSF.*

{__18} *Prepare the map of the GPS track so that the digital orthophoto will be properly positioned: Use the U.S. State Plane 1983 coordinate system, Kentucky North Zone (1601). The datum is NAD83 (Conus). The altitude units, coordinate units, and distance units are Survey Feet. The Time Zone is Eastern Daylight.*

{__19} *Bring up the digital orthophoto image as a background to the files you have selected: Add the file ___:\GPS2GIS\ByeByeSA\no_sa_ncr_doq.tif to the set of background files to be displayed. Make the main Pathfinder Office window occupy the full screen. Maximize the Map window. Zoom to Extents. Zoom up on the part of the red track that crosses the DOQ. Look at the GPS tracks that represent the part of the trip on the divided highway that took place during the transition from "SA on" to "SA off."*

{__20} *Now add NoSA_Moving.COR. (You can easily add files to those already displayed. In the Open Window the currently opened file names are shown overlaid with a gray tint. If you want to keep those files, hold down Ctrl while you select the new files with the mouse.)*

Notice three things:

1. *That differential correction greatly improved the GPS track while SA was still on,*
2. *That differential correction improved the GPS track after SA was turned off. Now the track is not only in the highway right of way, but in the correct lane as well. And, just FYI,*
3. *When you open two files together Pathfinder Office connects the last fix of the first with the first fix of the second—which explains that red line that cuts cross country.*

{__21} *Zoom up more so you can get to see the actual lanes of the highway toward the eastern side. Look at the corrected trace. It appears that the car making the GPS track sideswiped other cars on the highway. (Fortunately, those cars had been there many months before so there was no problem.)*

ArcGIS

In this exercise you will look at two files generated by a car going around a portion of New Circle Road and returning to the starting point, in September of 1999 (9/99). Most of the trip took place using the four-lane divided highway.

You will recall from Chapter 3 that a digital orthophoto image is an aerial photograph that has been calibrated so that it may be used as a map. Here again we make use of this capability. You will look at a DOQ of part of that highway and see the GPS tracks overlaid on it.

{__22} *Start ArcMap. Navigate to IGPSwArcGIS\CircleRd_999. Add as data j4lse9.tif. This will provide an image and it will also set the Data Frame Properties to be*

NAD_1983_StatePlane_Kentucky_North_FIPS_1601_Feet

Check this by right clicking the data frame and choosing Coordinate System. Cancel the Data Frame Properties window.

The files of uncorrected GPS data are S090519A and B. Add S090519A_ uncorrected.shp as data. You may get a warning that the file's coordinate system is different from that of the data frame. S090519A_uncorrected.shp is in WGS 84 latitude and longitude coordinates. However, if you proceed (do) ArcMap will project the shapefile on the fly. But don't ever let this capability fool you into thinking that the underlying coordinate system has been changed. If you wanted to convert the coordinate system of a shapefile you would use ArcToolbox > Data Management Tools > Projections and Transformations > Feature > Project.

{__23} *Add the return trip, S090519B_uncorrected.shp. Make both of these lines green. Zoom to Full Extent.*

{__24} ***Examine the files:*** *You will notice an area at the westernmost area of the trace where the two tracks run along together along an arc. If you zoom up on that area you will be able to see that the tracks merge and cross—an indication that the GPS points contained considerable error. Return the window to the full extent of the GPS tracks.*

{__25} *Using Zoom In, enlarge the orthophoto image. Notice how the GPS tracks across it follow the divided limited access highway (sort of). However, they appear to be well outside the lanes of the highway.*

{__26} *Now open S090519A_corrected.shp and S090519B_uncorrected.shp. You should see a considerable improvement: the tracks remain separated for their entire distance (except at the beginning and end where sharp turns and two-lane roads were involved.) If you zoom to the layer of the orthophoto you will see a dramatic improvement. It appears that the car actually stayed on the pavement.*

Differential Correction Also Improves GPS Accuracy with SA Off

Now we will revisit the Chapter 3 Project in which we looked at the effect of turning SA off as a car was traveling around New Circle Road. You will again see the file taken as SA went off, and then see even further improvement by the shapefile made by correcting that file. The proof will be shown by the DOQ we used in Chapter 3.

{__27} *Start ArcMap. Add as data the TIF file no_sa_ncr_doq.*

{__28} *Display NoSA_Moving_Lat_Lon.shp. Zoom up on the northeast area of the DOQ. Notice that while the GPS trace is in the right of way of the road it is in the wrong lane. Zoom to the extent of the DOQ layer. Then zoom to the northern half of the DOQ.*

{__29} *Now add NoSA_Moving_corrected.shp. Notice two things:*

1. *That differential correction greatly improved the GPS track while SA was still on,*
2. *That differential correction improved the GPS track after SA was turned off. Now the track is not only in the highway right of way, but in the correct lane as well.*

{__30} *Zoom up more so you can get to see the actual lanes of the highway toward the eastern side. Look at the corrected trace. It appears that the car making the GPS track sideswiped other cars on the highway. (Fortunately, those cars had been there many months before so there was no problem.)*

PROJECT 4G—THE UK CAMPUS

More On GPS Files and DOQs

Pathfinder Office

In this Project we verify again that differential correction has great value—here we superimpose a GPS trace made by walking, on two digital orthophotos—one of much higher resolution than the other.

{__1} *Make a Pathfinder Office project named UK_Campus_yis. Specify its folder to be ___:\GPS2GIS\UK_Campus_yis.*

{__2} *Use the operating system to copy the contents of*
___:\GPS2GIS\UK_Campus
to
___:\GPS2GIS\UK_Campus_yis
by doing the following:

Minimize the entire Pathfinder Office window. Open two windows showing the contents of both folders, UK_Campus and UK_Campus_yis. The latter should be empty except for folders labeled "backup," "base," and "export." Click in the title bar of UK_Campus. Hold Ctrl and press the "A" key to select all folders and files. Hold down the Ctrl key and press the "C" key to copy the highlighted files and folders onto the operating system clipboard. Click the title bar of UK_Campus_yis, hold down Ctrl, and press the "V" key to paste the contents of the clipboard into that folder. Say Yes to any complaints that folders already exist.

The folder ___:\GPS2GIS\UK_Campus_yis should now contain at least the following:

- the rover file r020919b.ssf
- the doqqqq image file j41se8w.tif and its world file j41se8w.tfw
- a folder "base" that contains U0020919.exe and L0020919.ssf
- two empty folders, "export"* and "backup"

The file r020919b.ssf was made by a student carrying a GPS receiver on the University of Kentucky campus, from a point almost on South Limestone Street, along a sidewalk that traverses the front lawn (which contains the McVey monument), past the Administration building, across the southwest border of the Plaza in front of the main faculty office building, along a sidewalk between two major library buildings (which blocked the signal in that area), and past the president's home. He turned right on Rose Street, traveling southwest until coming to a main UK entrance where he turned right again into a parking area. Another right turn led him to intersect his earlier path. (If you are interested in an Internet map of the place, the coordinates are those of the McVey monument: 38*02'20.57581"N and 84*30'18.67673"W.)

The files j41se8w.tif & j41se8w.tfw constitute a digital orthophoto of a portion of the UK campus.

The files in the folder "base" (U0020919.exe and L0020919.ssf) are files downloaded from two Trimble Community Base Stations in Lexington. Either of them could be used to adjust the points in r020919b.ssf to make r020919b.cor.

{__3} *Fix up the parameters: In Pathfinder Office, under Options, make the Distance units U.S. Survey Feet. For Coordinate System pick U.S. State Plane 1983. The Zone should be Kentucky North 1601. Coordinate Units should be U.S. Survey Feet; the Altitude Units should be Survey Feet as well. Make the Time Zone Eastern Std USA.*

{__4} *Open and display the waypoint file McVey_yis.wpt file. (You may have to browse for it. It's in ___:\GPS2GIS\McVey_NGS. You may also have to zoom to the full extent of the open feature and background files. Close any besides the waypoint.)*

---

* We'll use the export folder in Chapter 6 when we make a shapefile from the data in r020919b.cor.

{__5} *Display the file j41se8w.tif in the background (after removing any other files in the Load Background Files window).*

{__6} *Bring up the Map and make it full screen. Open the Position Properties window. The Map Window may shrink. Resize the Map window and move the Position Properties window off to the side so that both windows are visible and not overlapping. Zoom to Extents. Open and display the rover file r020919b. ssf.*

{__7} *Make the Not In Feature layer be displayed with heavy red dots; the dots should not be joined.*

{__8} *Zoom up on the beginning of GPS track. It is pretty hard to see what the path was. Using the Position Properties window, look at the location of the First fix. Click the DOPs tab. Move several positions along the track using the > button. Note that when the path takes a sudden jump a different set of "Satellites" is used to compute the position. This shows that the error produced by one set of satellites may be quite different than the error produced by another set.*

{__9} *Note the position of the McVey monument relative to the GPS track.*

Correct the Rover File

{__10} *Click on the button on the Start bar that will bring up the window of ___:\GPS2GIS\Campus_Walk_yis. (Alternatively you could click on the project folder on the Project toolbar.) Pounce on the "base" folder to bring up the list of files in ___:\GPS2GIS\UK_Campus_yis\base. Pounce on U0020919. exe to have the computer execute it. You should shortly see the appearance of U0020919.ssf. The self-extracting (.exe) base file just created an .ssf file. Bring back Pathfinder Office if you minimized it.*

{__11} ***Differentially correct r020919b.ssf using U0020919.ssf:*** *The only new wrinkle here is, since you have two base station files, you can ask Pathfinder Office to choose one of them based on the Preferred Base File Prefix. Choose U. Then when you get the screen showing the Reference Position of the base station, write down the Latitude, Longitude, and Altitude here:*

Lat: _____ Lon: _____ Alt: _____

Complete the process, making sure the resulting COR file goes into ___:\ GPS2GIS\UK_Campus_yis.

{__12} *Display r020919b.cor. It should look a lot more reasonable. Compare the path with the text description of the GPS track given above. Use zoom and*

FIGURE 4.13 Apparent "displacement" of top of building because orthophotos show correct position only at ground level.

pan (including the Auto-pan to Selection feature if you want to move along the path using the Position Properties window) to look at how closely the track follows the sidewalks.

(This is a good place to notice a characteristic of orthophotographs. Early in the walk, as the GPS track crosses a plaza, you will notice a tall building to the northeast of the track. Zoom up on that building. You can see the southwest side of the building clearly. You may assume that, in reality, the top of the building is precisely over the base, though on the photo the top of this 18-story building appears to be displaced by some 80 feet horizontally. (Use Measure.) The lesson here is that **digital orthophotos depict locations accurately only at ground level**. See Figure 4.13.

{__13} *Use the operating system to copy (Ctrl-C) U0020919.SSF that is in ___:\GPS2GIS\UK_Campus_yis\base and paste (Ctrl-V) it into ___:\GPS2GIS\UK_Campus_yis.*

{__14} *Open U0020919.SSF (in the Base folder of ___:\GPS2GIS\UK_Campus_yis) as a Background file. Even though it is a base station file it may be displayed (and differentially corrected as you will see shortly). Zoom to the full extent of all open and background files. Under View ~ Layers ~ Background display Not in Feature files with a thinnish pink line. The building over which the fixes cluster is the Forestry Building, where the UK Community Base Station is located. Zoom in so that you can see this building and those surrounding it. Measure the span of the fixes in this file: _____ This will give you a good idea of the best accuracy that could be expected when selective availability was active.*

Correct a Base Station File

{__15} *Start the process of differentially correcting U0020919.ssf with L0020919.ssf (both in the "base" folder) to produce U0020919.cor in the ___:\GPS2GIS UK_Campus_yis folder: In the Differential Correction window choose Settings to make certain that only corrected positions are placed in the output file. (Make sure that you specify the preferred base station file letter to be "L," not "U"; correcting a base station file with itself would be sort of cheating.) When finished with the differential correction process, display this file. Note that the points all fall within a circle of diameter of less than a foot, except for one outlier that pushes the span to about 1.3 feet. You will have to zoom out considerably to see the context of the fixes.*

This exercise gives you an idea of how well differential correction can work. Basically, for using GPS for GIS, where accuracy is all important, differential correction is vital. It is therefore relatively unimportant that selective availability was removed. It is hard to imagine a situation in which you would not differentially correct GPS files that are going to be used to build a spatial database.

{__16} *Use the grouping utility to make the file U0020919_pt.cor.*

{__17} *In the ___:\GPS2GIS\UK_Campus_yis folder you will find a waypoint file named UK_Community_base_Station.wpt. It consists of a waypoint named Calvin. Open that file and display the waypoint.*

{__18} *Repeat the process of correcting a base station file, but this time correct a file that was taken after selective availability was eliminated. Correct U0050219.SSF with L0050219.SSF.*

{__19} *Change settings: Lat-Lon, WGS 1984, ddmmssss, meters all around, and HAE. By the way, changing the projection will force Pathfinder Office to unload the DOQ, since image files cannot be projected and maintain their integrity.*

{__20} *Use the record editor to derive statistics for the file U0020919. Compare the results with the location of the base station that you wrote down earlier.*

Surveyed: Lat: _____ Lon: _____ Alt: _____
Calculated: Lat: _____ Lon: _____ Alt: _____
Differences: _____ _____ _____

{__21} *Move to the parent folder: ___:\GPS2GIS\UK_Campus_yis. Delete the .tif file (for disk space reasons).*

When one does a set of scientific experiments to determine the truth about "something" one usually wants to hold fixed ("constant") most of the things that could affect the outcome of the experiments while varying one or two other things. For example, if you were trying to determine a law for how long it took an object dropped from a height to hit the ground you might want to drop the object from different heights and record the duration of the fall. But you would always want to use the same object—not a billiard ball one time and a Ping-Pong ball the next. One of the problems in determining accuracies of GPS measurements is that it is impossible to repeat experiments under exactly the same conditions. Everything changes. The satellites move. The atmosphere changes composition. And so on. So absolute statements about GPS accuracy are hard to make honestly. One approach to determining GPS accuracy is to make lots of different measurements at different times and attempt to "get an idea" of "what you might expect." Not completely satisfying but it's the best we can do.

What follows is a mini-experiment, easily done, that will give you an idea of the best you could expect from differential correction. We take five base station files and correct them (with another base station) and look at how much improvement was made. Base station files are good SSF files to work with since (a) we know exactly where the antennas are, (b) they collect a lot of data (every three or five seconds), and (c) their antennas are mounted so that they get really good reception.

{__22} *Make a project that points to the folder ___:\GPS2GIS\Compare_ ssf_cor. There you will find SSF files U0070707, U0080808, U0090910,* U0101010, and U0111111. These are files generated since selective availability went off. They are about a month apart so they used different constellations. While the file names have a certain pattern to them the files themselves probably constitute a fairly random sample of DOPs and signal strengths. In the "base" folder are base station files that will correct most of the points of the SSF files.*

{__23} ***Differentially correct the five files.*** *You can do all five files at once. Start the differential correction process. Select the first SSF file by clicking on its name, then select the remaining four SSF files by holding down Ctrl and clicking. Ignore complaints that the files are not rover files. Conduct a Local Search for the base files (preferred base station prefix "L") and you will see all five files set up to be corrected at once. Approve everything and wait—there are lots of points for the computer to calculate.*

---

* I wanted to use U090909, for the sake of maintaining the pattern of the file names, but the equivalent base station file wasn't available. That happens sometimes. For critical work you probably should investigate ahead of time the reliability of any base station you intend to use.

{__24} *Group each of the SSF files into five point features. You can again process all five files at the same time. Put the results into GROUPED.SSF.*

{__25} *Open GROUPED.SSF. Represent the waypoint Calvin with a blue triangle. Represent a Point_generic feature as a red square. You may have to Zoom to Extents to get the waypoint included on the map. Measure some distances. This gives you a good idea of the best horizontal accuracy you can expect from uncorrected data.*

{__} *Open the position properties window. Click on the waypoint and note it's altitude (HAE). _____. Now click on the other points and note their altitudes. (You can tell which is which by the date.) You now have an idea of the best vertical accuracy you can get from uncorrected data.*

{__26} *Group each of the COR files into GROUPED.COR. Display and measure these. Note their altitudes.*

{__27} *Now open and display both GROUPED.SSF and GROUPED.COR together. You should now have a pretty good idea of how much differential correction can improve excellent SSF files. In general, your SSF files will not be as good—so take that into consideration.*

Look at a High-Resolution Color DOQ

Sometimes, as you've seen, DOQs are in black, white, and grayscale. In this project you will see the results of part of the walk across the University of Kentucky campus overlaid on a high-resolution color orthophoto. The grayscale orthophotos you used previously have cells a little more than 3 feet on a side. The color orthophoto that you will load next uses 6-inch cells.

(Unless you have a computer with a fairly sizable memory and a good bit of free disk space, you may not be able to do this part of the project. The size of the grayscale orthophotos is somewhat over two megabytes. The color orthophoto covers somewhat less territory but has resolution about 6 times higher (which means about 36 times as many cells in a given area, and—because it is in color—requires many more bits per cell. Therefore, it will not surprise you that the color orthophoto is larger in terms of memory size. It may surprise you that it is about 50 times larger: 102 megabytes compared to about 2.2 megabytes.)

{__28} *Change settings: U.S. Survey Feet everywhere. U.S. State Plane 1983. Kentucky North 1601. Eastern Standard.*

You may proceed in either of two ways, depending on the speed of your CD-ROM drive and disk space available. One is to load the large color orthophotos into PFO directly from the CD-ROM; the other is to copy the data onto the hard drive and load the images from there. If you choose the latter:

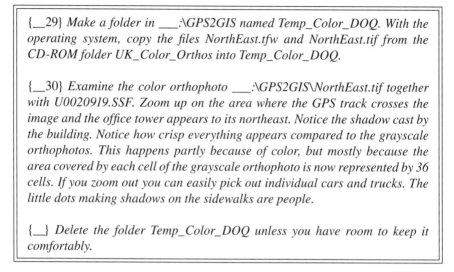

{_29} *Make a folder in ___:\GPS2GIS named Temp_Color_DOQ. With the operating system, copy the files NorthEast.tfw and NorthEast.tif from the CD-ROM folder UK_Color_Orthos into Temp_Color_DOQ.*

{_30} *Examine the color orthophoto ___:\GPS2GIS\NorthEast.tif together with U0020919.SSF. Zoom up on the area where the GPS track crosses the image and the office tower appears to its northeast. Notice the shadow cast by the building. Notice how crisp everything appears compared to the grayscale orthophotos. This happens partly because of color, but mostly because the area covered by each cell of the grayscale orthophoto is now represented by 36 cells. If you zoom out you can easily pick out individual cars and trucks. The little dots making shadows on the sidewalks are people.*

{_} *Delete the folder Temp_Color_DOQ unless you have room to keep it comfortably.*

ArcGIS

In this Project we verify again that differential correction has great value—here we superimpose a GPS trace made by walking, on two digital orthophotos—one of much higher resolution than the other.

{_31} *Start ArcMap. Begin to add data from __:\IGPSwArcGIS\UK_Campus.*

The folder should contain at least the following:

- the doqqqq image file j41se8w.tif and its world file j41se8w.tfw
- the GPS-based shapefile r020919b_uncorrected.shp
- the differentially corrected version of r020919b

The file r020919b_uncorrected.shp was made by a student carrying a GPS receiver on the University of Kentucky campus, from a point almost on South Limestone Street, along a sidewalk that traverses the front lawn, past the Administration building, across the southwest border of the Plaza in front of the main faculty office building, along a sidewalk between two major library buildings (which blocked the signal in that area), and past the president's home. He turned right on Rose Street, traveling southwest until coming to a main UK entrance where he turned right again into a parking area. Another right turn led him to intersect his earlier path. (If you are interested in an Internet map of the place the coordinates are: 38°02'20.57581"N and 84°30'18.67673"W.)

The files j41se8w.tif and j41se8w.tfw constitute a digital orthophoto of a portion of the UK campus.

{__32} *Add as data j41se8w.tif. The Coordinate System is U.S. State Plane 1983. The Zone is Kentucky North 1601. Coordinate Units are U.S. Survey Feet; the Altitude Units should be Survey Feet as well.*

{__33} *Add as data the rover file r020919b_uncorrected.shp. Make the track a red color.*

{__34} *Display r020919b_corrected.shp. It should look a lot more reasonable. Compare the path with the description of the GPS track given above. Use zoom and pan to look at how closely the track follows the sidewalks.*

(This is a good place to notice a characteristic of orthophotographs. Early in the walk, as the GPS track crosses a plaza, you will notice a tall building to the northeast of the track. Zoom up on that building. You can see the southwest side of the building clearly. You may assume that, in reality, the top of the building is precisely over the base, though on the photo the top of this 18-story building appears to be displaced by some 80 to 90 feet horizontally. (Use Measure.) The lesson here is that **digital orthophotos depict locations accurately only at ground level**. See Figure 4.13.

When one does a set of scientific experiments to determine the truth about "something" one usually wants to hold fixed ("constant") most of the things that could affect the outcome of the experiments while varying one or two other things. For example, if you were trying to determine a law for how long it took an object dropped from a height to hit the ground you might want to drop the object from different heights and record the duration of the fall. But you would always want to use the same object—not a billiard ball one time and a Ping-Pong ball the next. One of the problems in determining accuracies of GPS measurements is that it is impossible to repeat experiments under exactly the same conditions. Everything changes. The satellites move. The atmosphere changes composition. And so on. So absolute statements about GPS accuracy are hard to make honestly. One approach to determining GPS accuracy is to make lots of different measurements at different times and attempt to "get an idea" of "what you might expect." Not completely satisfying but it's the best we can do.

Look at a High-Resolution Color DOQ

Sometimes, as you've seen, DOQs are in black, white, and grayscale. In this project you will see the results of part of the walk across the University of Kentucky campus overlaid on a high-resolution color orthophoto. The grayscale orthophotos you used previously have cells a little more than 3 feet on a side. The color orthophoto that you will load next uses 6-inch cells.

(Unless you have a computer with a fairly sizable memory and a good bit of free disk space, you may not be able to do this part of the project. The size of the grayscale orthophotos is somewhat over 2 megabytes. The color orthophoto covers somewhat less territory but has resolution about 6 times higher (which means about 36 times as many cells in a given area, and—because it is in color—requires many more bits per cell. Therefore, it will not surprise you that the color orthophoto is larger in terms of memory size. It may surprise you that it is about 50 times larger: 102 megabytes compared to about 2.2 megabytes.)

You may proceed in either of two ways, depending on the speed of your CD-ROM drive and disk space available. One is to load the large color orthophotos directly from the CD-ROM; the other is to copy the data onto the hard drive and load the images from there. If you choose the latter:

{__35} *Make a folder in ___:\MyGPS_yis named Temp_Color_DOQ. With the operating system, copy the files NorthEast.tfw and NorthEast.tif from the CD-ROM folder GPS2GIS\UK_Color_Orthos into Temp_Color_DOQ.*

{__36} *Add as data ___:\MyGPS\Temp_Color_DOQ\NorthEast.tif. Zoom to the layer. You will see the GPS tracks in the southwest corner.*

{__37} *Examine the color orthophoto. Zoom up on the area where the GPS track crosses the image and the office tower appears to its northeast. Notice the shadow cast by the building. Notice how crisp everything appears compared to the grayscale orthophotos. This happens partly because of color, but mostly because the area covered by each cell of the grayscale orthophoto is now represented by 36 cells. You can easily pick out individual cars and trucks. The little dots making shadows on the sidewalks are people!*

{__38} *Delete the folder Temp_Color_DOQ unless you have room to keep it comfortably.*

Project 4H—DOP Matters

You must not assume that differential correction will take out all errors, or even all significant errors. In particular multipath errors and errors caused by high PDOP can remain after the correction procedure has been applied.

Pathfinder Office: The file H092702B.SSF is an 11-hour file taken by a GPS receiver with the PDOP mask set to 99. File H092720B.COR is a file made by correcting 4 hours of the data of file H092720B.SSF.

{__1} *In the folder ___:\GPS2GIS\High_PDOP open the file H092720B.SSF. Display it, joining its points with a thin black like. Note that, even though selective availability was not active, some points are very wide of the mark— as much as 200 meters. Measure the distance in meters from the center of the cluster to the four fixes that are farthest away. ____ ____ ____ ____*

{__2} *Check on the DOP values to the most far ranging points at the ends of spikes. Some have very high PDOP values. Others, probably the result of multipath error, have reasonable PDOPs.*

{__3} *Open file H092720B.COR as a background file (it's in the same folder). Set up View ~ Layers ~ Background so that Not in Feature files display as a thin red line. Note that the spikes are still there. In a couple of cases it appears that differential correction made things worse.*

The moral of the story is that differential correction cannot compensate for errors caused by high PDOP or multipath.

ArcGIS: The file H092702B_uncorrected.shp is an 11-hour file taken by a GPS receiver with the PDOP mask set to 99. File H092702B_corrected.shp is a file made by correcting 4 hours of the data of file H092702B_uncorrected.shp.

{__4} *Start ArcMap. In the folder ___:\IGPSwArcGIS\High_PDOP open the file H092702B_uncorrected.shp. Display it, using a black point. Use the Identify tool to look at the points that lie a long way from the cluster. Particularly observe the Max_PDOP value. Some have very high PDOP values. Others, probably the result of multipath error, have reasonable PDOPs. Note that, even though selective availability was not active, some points are very wide of the mark— as much as 200 meters. Measure the distance in meters from the center of the cluster to the four fixes that are farthest away. ____ ____ ____ ____*

{__5} *Add the shapefile H092702B_corrected.shp. Display it, using a red point. Note that the outliers are still there. In a couple of cases it appears that differential correction made things worse.*

The moral of the story is that differential correction cannot compensate for errors caused by high PDOP or multipath.

PROJECT 4I—LOOKING AT GPS FIXED IN THE CONTEXT OF PUBLISHED DATA

ArcGIS: The following Project is done in ArcGIS with ArcCatalog and ArcMap.

ESRI Data and Maps

In addition to providing GIS software ESRI also releases large quantities of spatial data—five DVDs containing highly compressed spatial data sets—that come with ArcGIS. ESRI describes the data sets as follows:

> Data & Maps for 2008 is organized on five DVDs. The Data & Maps and StreetMap North America DVD contains the StreetMap North America data set as well as the Data & Maps vector data. The other four DVDs include all the elevation and image data sets. One of the four contains the 90-meter Shuttle Radar Topography Mission (SRTM) global digital elevation model with all the void areas filled as well as many other world-wide elevation and image data sets. The remaining three DVDs contain regional data for North and South America, Europe and Africa, and Asia and Australia. Each of these regional DVDs includes global imagery captured at 150-meter resolution as well as shaded relief derived from the SRTM global digital elevation model. In addition, all five DVDs contain the Data & Maps HTML-based Help system.

The Data & Maps vector data include census spatial data such as roads, lakes, rivers, counties, and so on. We will take a look at a tiny fraction of what is available and put it into the context of a GPS track. It is worth reminding you that GIS data sets have no associated scale. Scale is a concept that relates that which is on the ground to that which is on a piece of paper or a computer screen. GIS data sets are associated with actual, real world distances. You only need to concern yourself with "scale" when you plot out GIS data sets, or view them on a monitor screen.

ESRI Data and Maps is a good data set to illustrate principles but if you want the most current national or international geographic data you should explore several sources. One really good place to start is on the Internet: http://geographynetwork. com contains a plentitude of free data, government data, and commercial data.

We now turn to examining GPS data together with GIS data from ESRI Data and Maps. In this project, you will bring up a GPS track of a vehicle that traveled from Lexington, Kentucky, south on Interstate 75, west around Knoxville, Tennessee, to a bank of Fort Loudon Lake (the Tennessee River) near Fort Loudon Dam. You will see this track in the context of the counties of Kentucky and Tennessee, major roads and hydrological features. The GPS file that was taken was named I090317A.SSF. From that file a shapefile of the same name* was generated, using latitude and longitude in decimal degrees.

ArcMap and ESRI Data

{__} *Start ArcMap. In the folder* ___:\IGPSwArcGIS\Lex_Knox, *find and add the following shapefiles in this order:*

1. *ESRI_County_Data.shp*
2. *I090317A.shp*
3. *Interstate-75.shp*

---

* The Trimble software generates shapefiles of the name posnpnt.shp rather than the name of the GPS input file (e.g., I090317A). Once in ArcGIS the name can be changed back to the original file name or some other name.

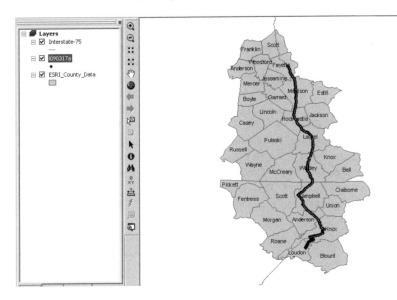

FIGURE 4.14 GPS track and counties named using labels.

{__2} *Make I090317A.shp black dots of Size 4. Make Interstate-75 a red line of Width 1. Show ESRI_County_Data with a fill color of light yellow. Make sure that Interstate-75 layer name is at the top of the Table of Contents.*

You should see a number of counties that contain a portion of Interstate 75; the GPS track is composed of dots that you can see under the red line of the Interstate. Label the counties with NAME. (Right click on the ESRI_County_Data and pick Properties. In the Layer Properties window click Labels. Check the box "Label features in this layer" and make the Label field NAME.) See Figure 4.14.

Projecting Coordinates

Had the Earth been formed as a cube instead of a quasi-sphere many things regarding representation of locations and features on it would be simpler. (Of course, many other things would be more complicated, but here we are just concerned with methods of representation of features on the Earth's surface.) The coordinate system that works best to represent a single location on Earth is a spherical coordinate system. It can be applied perfectly, producing no distortion. Unfortunately, if you have to deal with, say, two point locations, the mathematical formulas for determining the distance between them, or the direction one must travel to get from one to the other, are complex. For many reasons, it is useful to represent areas of the surface of the Earth on a flat sheet (or computer screen) in a Cartesian coordinate system. The distance and area formulas are simple; the medium of display (paper) is convenient. (A model globe, which could provide theoretically complete accuracy of representation, distance, shape, and size, would have to be astoundingly large to provide any reasonable level of detail for human-scale activity.) When you make a transformation (called a projection) from a spherical system to a Cartesian one, you inevitably introduce inaccuracies in the locations of points that were correct in the spherical coordinate system. So when dealing with large areas of real estate you are always caught in the dilemma of choosing (1) exact representation (giving up representative plotting on a flat surface and easy distance calculation, or (2) inexact representation (projection) which allows easy calculation and convenient display.

The image in the View—projected onto the flat screen of your monitor—is in latitude-longitude decimal degrees. Such images are distorted because nowhere on Earth (not even at the equator, where it is close) does a degree of latitude represent the same distance as a degree of longitude. You cannot tell that this GPS track is distorted, because it is basically a line running from north to south. But when we put other data with it the distortion becomes apparent. You will therefore change the view so that it shows data in Universal Transverse Mercator (UTM) coordinates. In UTM there is slight inaccuracy and very slight distortion, but even these effects are much reduced over small areas.

{__3} *Make certain the mouse pointer is a simple arrow by choosing the arrow symbol (it is on the Tools toolbar). Slide the cursor over the map. Notice that the Status Bar gives the coordinates in Decimal Degrees. Place it over the north end of the GPS track. Note the coordinate values, reported on the Status Bar. The longitude (the east-west coordinate) should be something like −84.43.* What is the latitude coordinate?* _____

To reduce the distortion, you will convert the picture of the data you see from the spherical coordinate system (the lat-lon graticule based on degrees and decimal fractions of a degree) to the UTM rectangular coordinate system for Zone 16. The UTM units will be meters. We will actually only make one new shapefile (based on counties) in the UTM system. Then we will rely on ArcMap to project the others on the fly.

{__4} *Close ArcMap without saving the existing map. Start ArcCatalog. Open ArcToolbox (by clicking on the red toolbox icon). In ArcToolbox navigate to*

   Data Management Tools > Projections and Transformations > Feature > Project.

Double click the Project[†] tool. For the Input navigate to

   IGPSwArcGIS\Lex_Knox and double click ESRI_County_Data.shp.

Change the name of the output by substituting the letters UTM for Project, so that the full name is

   IGPSwArcGIS\Lex_Knox\ESRI_County_Data_UTM.shp.

Click the icon to the right of Output Coordinate System. Click Select in the Spatial Reference Properties window. Navigate:

   Projected Coordinate Systems > Utm > Wgs 1984 > WGS 1984 UTM Zone 16N.prj

---

* This western hemisphere longitude designation may also be represented as 84°W (for West) in other systems and software. The only problem occurs when you leave off both the minus sign and the W (or use both of them—which makes it East again).

† Recall that I said that the word "project" has two meanings. Well, here is the other one—a verb.

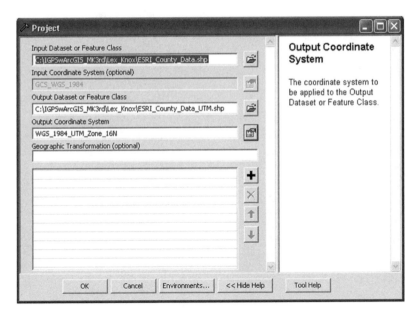

FIGURE 4.15   Tool for projecting lat-lon coordinates to UTM coordinates.

Click Add. Read the information about this projection. Note that the Central Meridian (the center north-south line that bisects the zone) of the zone is 87 degrees west (−87°). Click Apply. Click OK.

{__5} The Project window should look like Figure 4.15. If so, click OK. When the projection is complete click Close.

{__6} In ArcCatalog, navigate to IGPSwArcGIS. From the View menu click Refresh. Click Preview at the top of the screen. Expand Lex_Knox. Click ESRI_County_Data. shp and examine it. Now click ESRI_County_Data_UTM.shp. Notice the difference. Flip back and forth between the two representations, which are shown as Figure 4.16. Note that the UTM representation does not contain the distortion engendered by plotting the longitude and latitude coordinates (of the Counties_DD) directly on a Cartesian grid (which is what your computer screen is). The UTM representation is the more realistic one. That is, if you could delineate the county boundaries with, say, visible yellow plastic tape on the surface of the Earth and view the entire area from an airplane at high altitude, the area would look like the UTM version.

{__7} Start ArcMap. Add as data ESRI_County_Data_UTM.shp. Notice that the Status bar now shows the coordinates in Meters. Add in the GPS track: I090317A. shp. Finally add in Interstate-75.shp.

{__8} Make I090317A.shp black dots of Size 4. Make Interstate-75 a red line of Width 1. Make ESRI_County_Data with a fill color of light yellow.

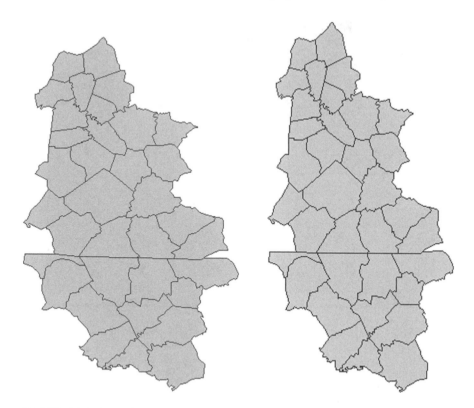

FIGURE 4.16   Areas in distorted lat-lon coordinates compared with (nearly correct) UTM coordinates.

{__9} *Label each county with its name.*

{__10} *Again slide the pointer around the view. Note that the coordinates at the north end of the GPS track are about 725,000 for the horizontal (that makes it 275,000 meters east of the central meridian of negative 87° (defined to be 500,000, right?) and 4,210,000 meters north of the equator for the vertical.* **It is important to understand that the original GPS track data set has not been converted—it is still in decimal degrees—but its visual representation has been altered to reflect the UTM coordinate system, because the initial data added to ArcMap was the new counties that you did reproject into UTM.**

Identifying and Selecting Particular Features of a Particular Shapefile

{__11} *Start the Identify tool. Move the pointer into the county in which the GPS track starts. Click on the county. What is its name? _____ Notice all the information that is available about it. Verify that the FIPS (Federal Information Processing Standard) code for Kentucky is 21 and for the county 067. You learn that the overall FIPS code is 21067, which indicates, if you did not already know it, how an overall FIPS code is assigned. What is the area of the county? _____ square miles.*

{__12} *Click on the county at the end of the GPS track. Loudon County, Tennessee? Click on the county just northeast of Loudon. It should be Knox. Dismiss the Identify window. Open the attribute table of ESRI_County_Data_UTM. What is the State_ FIPS of Tennessee?* _____

{__13} *From the Selection drop-down menu pick Select By Attributes. For Layer choose ESRI_County_Data_UTM. Double click on "NAME" to bring it into the SELECT box. Single click the equals (=) sign. Click on Get Unique Values and double click "Knox." Apply. OK.*

{__14} *In the table click on Show Selected Records. How many Knox Counties are there?* _____ *Shorten the table vertically so you can see the map. Zoom to the counties layer. Pick out the selected counties. Examine and then dismiss the attribute table. From the Selection drop-down menu pick Clear Selected Features.*

Adding Water

{__15} *From IGPSwArcGIS\Lex_Knox add Rivers_DD. Make the lines that delineate the rivers deep blue of width 4. Zoom to the extent of this layer. These are all the major rivers in the United States, according to the ESRI Data. Open the attribute table. Click to the left of the File ID number (FID) to select some rivers. Look at them on the map. Using Options on the table window clear selections. Dismiss the table.*

{__16} *Zoom now to the extent of the Interstate-75 layer. This step and the one above may give you some insight into the ESRI Data. There is much, much, much more. And, in terms of hydrological and transportation data there are many other much more detailed databases from other sources.*

{__17} *Zoom way in on the very end of the GPS track. It appears from the image that the vehicle that carried the GPS antenna must have been amphibious: several points appear on the other side of the river. Use the measuring tool to determine the distance from the river to the southern most point on the GPS track. What is it? _____ meters. So it appears that the antenna was a third of a kilometer on the other side of the centerline of the river—which in actuality is a major lake (Fort Loudon) which has considerable width itself. In reality the vehicle stopped about 50 meters from the water. Where does the problem lie?*

The GPS track probably has its usual level of accuracy. (These are uncorrected fixes taken when SA was active so you can expect about 100-meter accuracy.) The imported data sets are less accurate, however. They are taken from 1 to 2 million scale maps. That would certainly more than account for the problem that we see with the GPS track and the water. The moral of the story: know the accuracy of the data you are working with. Don't make decisions based on spatial data without knowing what sort of accuracy is involved.

PROJECT 4J—SEEING GPS DATA IN THREE DIMENSIONS

> **ArcScene:** *If you have the ArcGIS 3D Extension and the ArcScene software you can do the following demonstration.*

GPS data sets are inherently four dimensional: three spatial dimensions and the temporal dimension. While virtually all GPS receivers can collect fixes with all four dimensions not all software that runs on GPS receivers is programmed to collect vertical (Z) coordinates. In particular, you have to augment ArcPad if you want it to collect the height attribute. TerraSync does collect data fixes with all four dimensions as coordinates, as the following project will demonstrate.

{__1} *Start ArcCatalog. In the Tools drop-down menu select Extensions and turn on 3D Analyst.*

Elmendorf_GPS_3D_cor_SP_N.shp is a three-dimensional GPS track created by walking up the 0.2 miles of Elmendorf Drive on the left side of the street and back down on the other side. It was collected with TerraSync, differentially corrected, exported to a WGS 1984 DD file, and then projected to Kentucky State Plane North Zone.

Elmendorf_Area_DOQ_SP_N.TIF is a grayscale raster data set (a Digital Orthophotoquad) which includes the area of Elmendorf Drive.

DEM_K42_KY_SP_N.img is a Digital Elevation Model (DEM) of the area that provides the heights of the terrain with "square posts" (about 30 feet on a side whose height at the top represents the elevation).

{__2} *Make a Folder Connection with IGPSwArcGIS\3D_GPS. If the C: drive (or whatever drive you use to store IGPSwArcGIS) is expanded, collapse it so you can see this folder connection. Click on Elmendorf_Area_DOQ_SP_N.TIF. In the Preview box at the bottom of the screen select 3D View from the drop-down menu.*

{__3} *With the cursor that appears which looks like* *maneuver the raster data set. The left mouse button rotates it. The right mouse button zooms it in and out. Convince yourself that it is a flat image.*

{__4} *Still in ArcCatalog, look at Elmendorf_GPS_3D_cor_SP_N.shp with the Preview still set at 3D View. Maneuver it around to convince yourself that the fixes are truly shown in three dimensions.*

{__5} *Start ArcMap. In the Tools drop-down menu select Extensions and turn on 3D Analyst. From the 3D Analyst toolbar start ArcScene. Add as data Elmendorf_Area_DOQ_SP_N.TIF and Elmendorf_GPS_3D_cor_SP_N.shp. Make the GPS fixes bright green. Maneuver the image of the two data sets.*

The obvious problem is that Elmendorf_GPS_3D_cor_SP_N is "floating" above the DOQ, which is at zero elevation. If you use Identify and click on one of the GPS fixes you will see that its GPS height is somewhere around 1000 feet in elevation. To solve this problem we will use a Digital Elevation Model (DEM) of the area to boost the z-coordinate for the Elmendorf_Area_DOQ_SP_N.TIF to the proper altitude. This is sometimes referred to as "draping" the DOQ over the DEM.

{__6} *Add as data: DEM_K42_KY_SP_N.img. What are the lowest and highest elevations it contains? _____ and _____. Right click on the name of Elmendorf_Area_DOQ_SP_N.TIF and choose properties. You will see a Base Heights tab in the resulting window. Click it. Press the radio button next to "Obtain heights for layer from surface" and, using the drop-down menu, choose the DEM. Apply. OK. We no longer need to see the DEM so turn it off. However, we will use its values, so do not remove it from ArcScene.*

{__7} *Now you see the GPS track congruent (sort of) with the DOQ. Look at the two data sets from different angles and levels of zoom. Use the Identify tool (set it to identify information about the GPS points) to determine the elevation of a fix at the bottom end of the road. _____ At the top end. _____*

{__8} *Maneuver the scene so you can look straight down on the GPS track and DOQ. (The zoom and the pan (the hand icon) will help.) (Note: some of the GPS points will have disappeared underground; you could see them by flipping the image and looking from the bottom; if you get confused just click on Full Extent to get back to a view you can recognize.) Use the Identify tool to determine the height of a GPS fix. Turn on the DEM. Determine the height of the DEM at that point also. (For some reason you may have to turn off the DOQ to let the Identify tool penetrate through to the DEM.) Explore various deviations between the heights provided by the DEM and those associated with the GPS fixes.*

You should come away from this project with one lesson and one reminder: (1) GPS data (depending on the software driving the receiver) can be three dimensional a far as coordinates (not just attributes) are concerned, and (2) the vertical GPS accuracy, corrected or not, is almost always worse than the horizontal.

Project 4K—"Must Know" Information about DOQs

The route from Lexington to Knoxville (Project 4-I) requires crossing the Kentucky River at Clays Ferry. Due to the volume of traffic, the two-car ferry was replaced some time ago; today a six-lane bridge—part of Interstate 75—handles the traffic. That bridge is several hundred feet above the river. A GPS track across the bridge is not congruent with the bridge span, as you will see if you do the following:

{__1} *Start ArcMap. In ___:\IGPSwArcGIS\Clays_Ferry\ you will find an image: CLAYS_FERRY.TIF. Add that as data. Maximize the map. Locate the bridge and zoom up on it, showing primarily the bridge and some highway at each end. Note that the bridge seems to bend toward the southwest.*

FIGURE 4.17   Displacement of bridge spans in orthophoto due to altitude differences.

{__2} *In the same folder you will also find Clays_Ferry.shp, which is the corrected GPS track of points representing fixes taken by a car crossing the bridge from the northwest to southeast. Add that as data as well. Make its legend a bright yellow square of size 8 and turn it on. The result should appear as in Figure 4.17. Notice that the track of GPS fixes starts off with the car not only on the Interstate but in the right lane. As the car enters the bridge, however, it seems to depart from the roadway, cross over to the northbound span, and then return to its original roadway. The data are accurate to 3 meters. Can you explain what the problem is? (Hint: you know something about DOQs and where they are, and are not, accurate.)*

PROJECT L—YOUR DATA II

Correct Your Own Data

In this project you will correct the data you took in Projects 2A (fixed point) and 2B (a trace made by moving the antenna). Or you might choose to correct some other data. You will learn how to find and download data from a base station.

Pathfinder Office

{__} *Return to the Project GPS2GIS\Data_yis. Examine the data you col-lected in Projects 2A and 2B. What are the last seven characters of the names of the base station files you would need to differentially correct these files?* Recall that the format of a base station file is XYMMDDHH.SSF.*

X_____

X_____

It may take a little research to determine if there is a base station within 300 miles of your location but GPS equipment vendors can usually supply the information. When you find one, fill out the form at the end of this chapter. If it's a Trimble compatible station you want what the company calls Community Base Stations. One way to find one is to point your Web browser at† http://www.trimble.com/trs/trslist.htm or http://www.trimble.com/gis/cbs/, and you can select base stations by state, province, region, and/or city. Information on how to retrieve the base station computer file can be found here also.

{__} *Bring up a browser and check out the URLs above. Explore your geo-graphic region. It may be obvious what base station is closest to you and how to download it. In any event, let's look at a different method before we do any downloading. Keep your computer connected to the Internet.*

A more elegant way to locate a nearby base station is provided directly by the Pathfinder Office software and a Trimble Navigation Web site. Once you open an SSF file that you want corrected, you can direct the software to auto-matically download a list of base stations to your computer, sorted in the order of distance from the positions in your SSF file. Assuming you have the correct permissions from the base station owner, the software can even download the base station file and automatically produce the COR file.

{__} *Start the differential correction utility. Select one of your data files as the Rover File. Under Base Files, click "Internet Search." In the window that comes up, click New. In the New Provider window choose "Copy the most up-to-date etc." option. What happens now may depend on how you are connected to the Internet, but what can happen is that your computer will download a list of many dozens of base stations. (See Figure 4.18.) The*

---

* Actually you might need more than two base station files. If your data collection extended over parts of two consecutive hours you need the base station files that cover the second hour as well.

† These worked at the time of this writing. But Web pages have a nasty habit of changing, so if these Web pages do not appear, go to http://www.trimble.com and navigate to, or search for, the "Community Base Stations" list or the "Trimble Reference Station List."

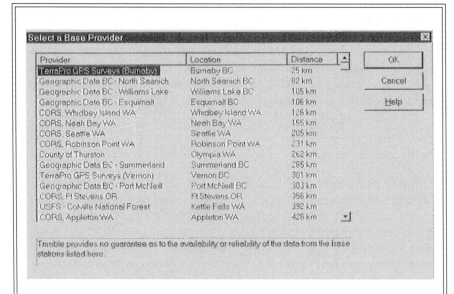

FIGURE 4.18   List of base stations from the Internet.

ones closest to your SSF file's location are listed first and, if you are on the North American continent, those in South Africa, Australia, and New Zealand are listed last. If you pick a base station and choose Properties, you have the option of going to their Web page directly or sending them e-mail. If you OK this window you may have the option of downloading the files you need directly. Or you may not, since there may be impediments— both technical and/or commercial.*

{__2} Using the information above, plus perhaps instructions from your teacher, colleagues, or employer, obtain the appropriate base station files. Such files are usually kept on hand at the base station locations for at least several months after they have been collected. As I've described, you may well be able to download these files from the Web, or through the download capability of your browser through a process called "anonymous FTP" (File Transfer Protocol). Bring the files into the base folder of the project you set up for the data in 2A and 2B. If the files are EXE files, or "zipped" files, convert them into SSF files.

{__3} Use the differential correction process on the files you collected. Display the SSF and COR files with Pathfinder Office software.

* Some stations charge for their services and differential correction files.

ArcPad: To get correct data out of ArcPad you basically have to collect it in corrected form using an SBAS such as WAAS. There do exist programs, such as Trimble Correct or Magellan GPS Differential, which will let you check data out of a GIS, correct it, and then put it back but these are beyond our scope here. Besides, these techniques are evolving as software evolves. You have the capability also to filter out positions which have not been SBAS corrected. This will work for collecting points but is of dubious value when collecting positions along a route, because you might miss a lot of fixes.

Exercises

Exercise 4-1

From reading and doing the projects in this chapter you have probably formed some ideas of the range of accuracies (both horizontal and vertical) you might expect from an autonomous GPS receiver. You also have some ideas of the ranges of accuracies you might expect once you differentially correct the data. Fill out the following table that puts these numbers together. (*Note*: There is no absolutely correct answer—only approximation of ranges.)

	Autonomous Best	**Autonomous Worst**	**Corrected Best**	**Corrected Worst**
Horizontal				
Vertical				

Pathfinder Office

Exercise 4-2: In \GPS2GIS\ROOFTOP you will find two files, C120600G.SSF and C120600G.COR, which represent another attempt to find the location of the coordinates of the rooftop of PROJECT 4A.

{__1} *Start the GPS software and set the Project to "Rooftop." Set up UTM coordinates in meters. Set the height as "Geoid," meaning height above mean sea level (MSL) and the altitude units as meters. Set the Datum to WGS-84.*

{__2} *Display these files so you can tell which is which (use the Grouping utility or the Background capability). Both plots look really terrible, with large dispersions. Use the Measure tool; you will determine distances between some points of more than half a kilometer. Further, the "corrected" file shows not one, but several clusters of points. What's going on here?*

{__3} *Use Position Properties to examine some points in each file. Particularly suspicious are the "Altitude" numbers. The altitudes all seem to be about 47.5 meters below sea level. Since the antenna was at a geographic point almost*

300 meters above sea level, we may suspect that some altitude setting is the culprit. We are also suspicious because the altitudes seem not to vary from point to point. Now notice that Position Properties reports that these points, whether corrected or uncorrected, are 2D rather than 3D.

So there is the answer. Some doofus (the author, actually) had the GPS receiver set on 2D or Auto 2D/3D while collecting this particular file. *You can only use those settings if you manually and correctly enter the altitude—* obviously not something that happened here, since the receiver thought the altitude was −47.5 meters MSL.

The moral of this story is an old one, related to using a computer to turn data into information. It is known as GIGO: "Garbage In, Garbage Out."

ArcGIS

Exercise 4-3: When you use GPS to collect line data you have to be aware that each fix on the line is a single reading, with the error implications that that has. In other words, there is no possibility of averaging. Something else to be aware of is that natural and human-made objects frequently are curvy and that a GPS linear track is composed of a sequence of fixes; when those fixes are connected the resulting lines may well not follow the object being represented.

{__4} *Start ArcMap. From __:\IGPSwArcGIS\FarmHouse add as data Pat_S_Points.shp. Make the dots bright green. Use Identify to determine which general direction the automobile that collected the fixes was traveling. _____*

{__5} *Add the TIFF file Pat_S.tif from the FarmHouse folder. Zoom in on the last 10 or so fixes of the GPS track. Note that these seem to be pretty much right on the country road—within 6 feet or so. If you measure the distance between the fixes you get about 200 feet, which was what the datalogger was set to receive.*

{__6} *Add as data Pat_S_Line.shp. Make the line bright red. You now get a somewhat different impression of the accuracy, as the red line goes slicing across fields and fence rows.*

So if you want to represent curvilinear features you (a) can't do it exactly, and (b) should use short distances between fixes. The distances can't be too short, however, because errors will be recorded and file sizes will become quite large.

Base Station Information

Base station name: _____

Base station location: _____

Approximate distance to the location of your files: _____

Web URL and/or FTP address: _____

Initial character of .SSF or .EXE files from this base station: _____

Contact Person & Telephone number: _____

Account name(s): _____

_____

Password: (record it elsewhere)

Procedure for downloading files:

_____

_____

_____

_____

Notes: _____

_____

_____

_____

_____

_____

5 Integrating GPS Data with GIS Data

In which you learn approaches to converting GPS-software* data files to GIS-software[†] data files, practice with sample GPS files and shapefiles, and then undertake the process on your own.

OVERVIEW

Another Datum Lesson: I teach at the University of Kentucky's Department of Geography, in Lexington. Several years ago I borrowed a GPS receiver from Trimble Navigation to use in a class. It had occurred to me that our departmental faculty directory listed names, office and home addresses, phone numbers, and so on, but not, as seemed appropriate for geographers, their geographic coordinates. So I sent my students out to park in my colleagues' driveways and collect GPS data, which we converted to an ESRI data set based on World Geodetic System of 1984 (WGS84) Universal Transverse Mercator (UTM) coordinates.

We then digitized the major roads from Lexington-area U.S. Geological Survey (USGS) 7.5-minute topographic quad maps based on the North American Datum of 1927 (NAD27). The maps indicated that to convert NAD27 to NAD83 (which is within centimeters of WGS84), one should move the grid lines 4 meters south and 6 meters west—hardly an issue for us, because our accuracy was limited by selective availability to perhaps 50 meters when calculated over a number of measurements. It was, therefore, distressing when a graphic overlay of these two coverages put the current departmental chair's house smack in the median of Lexington's limited-access beltway. (He, in fact, lives a couple of football fields north of there.)

What I learned, the embarrassing way, was that while the area's latitude and longitude coordinates were adjusted by a few meters in the 1927-to-1983 datum change, the UTM coordinates were adjusted by more than 200 meters in a north-south direction! So the statement on the map that the gridlines can be moved only a few meters was misleading—it referred only to the latitude and longitude grid lines. The UTM grid and the State Plane grid had much greater changes.

The moral of this story—which partly prompted the writing of this textbook—is that datum and coordinate issues can be tricky, and one must be especially careful when combining data from different sources.

---

* Trimble software, in particular.
[†] Environmental Systems Research Institute files, in particular.

195

REVIEWING WHAT YOU KNOW

You have been given, or you have collected, files with the extensions SSF (Standard Storage Format) or COR (for differentially CORrected) files. (If you have collected data with ESRI ArcPad, then the collected files are already in shapefile format.) These files have been processed and displayed using the Trimble Pathfinder Office software. Your goal now is to use these data files in a Geographic Information System (GIS) where they can be considered with many other data sources that you may have available. This is not a difficult process, but you really must be careful: it is quite easy to get what looks like a reasonable GIS file, but one that has the locations wrong, thus making the activity worse than useless.

The key to converting from Trimble files to ESRI files is to know that you need only convert from SSF (or COR) files to **shapefiles**. Once a shapefile has been obtained, the entire range of ESRI products is available for your use. It is also true that, no matter which Trimble products you use, ultimately you will have files in their standard file format. So it is as though you have a "data tunnel," with wide ranges of products on each side but with the restriction that the data must flow through a narrow passage. (See Figure 5.1.)

The process of making a shapefile is direct. Pathfinder Office generates a shapefile file (actually, three or more files in the same folder are needed to make an ArcView "shapefile") that is read directly by ArcGIS.

The reason to use a GIS with GPS data is so that you can combine locational data from a variety of sources. So, first and foremost, you must ascertain the parameters of the existing ESRI shapefiles into which you wish to integrate the GPS data. If you

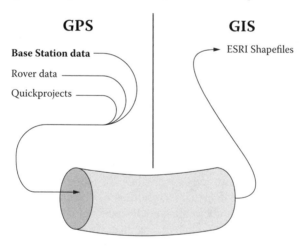

GPS

Base Station data

Rover data

Quickprojects

GIS

ESRI Shapefiles

GPS Data is converted with Trimble, Magellan, or other software perhaps using Microsoft ActiveSync.

Results can then be CAREFULLY combined (considering datum, projection, and units) with other ArcGIS data.

FIGURE 5.1   Data Tunnel: GPS data to ESRI products.

get this wrong, everything will be wrong thenceforth. Among the things you must consider are the following:

- Geodetic datum (for work in North America choices are usually NAD27, NAD83, WGS84)
- Projection, if any, used to convert the data from latitude-longitude representation to a Cartesian coordinate system
- Units of linear measure (e.g., meters, miles, survey feet, and many more)
- Units of angular measure (almost always degrees, but the issue of how fractional parts of a degree are represented can complicate things)

PRESCRIPTION FOR FAILURE: INCORRECT PARAMETERS

Datum

As you perhaps proved to yourself in Chapter 1, there may be a significant difference in the UTM coordinates of a point represented in NAD-27 and NAD-83 (WGS-84). In the United States, the values are different in the north-south direction about 200 meters, due primarily to humans learning more about the shape of their Earth. In the western United States the east-west difference can be around 100 meters (e.g., Seattle area: 93 meters; San Diego area: 79 meters); it is usually less in the eastern part of the country (e.g., Bangor, ME, area: 50 meters; Miami area: 17 meters; Lexington, KY, area: 2 meters). You need to be concerned that you convert the GPS file to the datum used by any shapefile you wish to combine with the GPS data. You may determine the datum of a data set in a variety of ways. Metadata, through ArcCatalog, is one way. Examining the properties of an ArcMap data frame is another—with the caveat that the data frame takes on the datum and projection of the first data set that is added to it when ArcMap is started; other data sets are visually, though not actually, converted to that first data set. If you are using shapefiles, the information may be present in a file with the extension "prj" appended to the name of the shapefile. A projection file contains information such as the following:

Projection	STATEPLANE
Zone	3976
Datum	NAD83
Zunits	NO
Units	FEET*
Spheroid	GRS1980
Xshift	0.00000000000
Yshift	0.00000000000
Parameters	

---

* ESRI software assumes that a "foot" is a "survey foot" but its projection engine can handle international feet. The difference is not insignificant over long distances. In 100 miles, for example, the difference is about a third of a meter—slightly more than a foot. Gunfire has been exchanged in disputes about property line positions differing by less than that. And the coordinate systems of many states span a few hundred miles.

You should probably carefully investigate the sources of the data and their processing history so you can be certain of the datum and other projection parameters used.

Projection

As you will recall, the 2D components of the Trimble SSF and COR files represent data in the latitude and longitude datum of WGS-84, though they may be displayed in other forms. A fundamental dilemma of a spatial analyst is that the most accurate way to depict a point on the Earth's surface is with latitude and longitude, but that the numbers that represent such a point are in a coordinate system (spherical) makes it harder to use these numbers in calculation for such quantities as distance and direction. Making these calculations in a projection is easier, but, of course, you get a (usually, slightly, if you are careful) wrong answer.

Further, maps that are shown in latitude-longitude are (usually) badly distorted visually, so most GIS users elect to store data in some projection in which the horizontal and vertical distances on the map correspond to the east-west and north-south distances on the Earth's surface. These maps appear (generally) much less distorted. However, there is now actual mathematical distortion for all but a few points. There is really no good solution to this dilemma; you must accept some inaccuracy. What's vital, however, is that you tell the Trimble conversion process the correct projection to use so that the inaccuracies in your GPS data will be consistent with those in the other data your are working with. ArcMap allows you to **display** data in another coordinate system besides the one in which it is recorded. Further, ArcToolBox contains routines which allow you to **convert** original data from one projection to another.

Linear Measurement Units

The choices are meters, feet (international), and survey feet. Survey feet formed the basis of the NAD-27 datum*; international feet were used in NAD-83 and WGS-84. What's the difference? Not much, but enough to be significant in some situations. The differences come from a slight distortion of each English unit to make it conform to the metric system. An international foot is based on the idea that there are exactly 0.0254 meters in an inch. A survey foot is based on the equality of exactly 39.37 inches and a meter. If these two conversions were equivalent, you should get exactly the pure, unit-less number one (1.0000000...) when you multiply them (0.0254 meters per inch times 39.37 inches per meter). They aren't and you won't. What is the product? Use a calculator. _____ The fractional part may look like an insignificant number, but it can mean a matter of several feet across a State Plane Coordinate zone.

Angular Measurement Format

Angular measurement units are important only if you are converting a file to a data set which uses the lat-lon graticule directly. If you do use the graticule, you will want to select degrees, and decimal fractions thereof, because ESRI products don't directly utilize minutes and seconds as coordinate values.

---

* State plane coordinate systems may also use survey feet.

The Old Conundrum: The "Spherical" Earth and the Flat Map

The GPS data in the receiver and in SSF files are stored in latitude, longitude, and height above ellipsoid coordinates. You can, of course, make ESRI shapefiles with latitude and longitude directly, as long as you select degrees and fractions of a degree as the output numbers, being sure to use enough decimal places. You may want to do this if you are going to combine the data with other data sets that are stored in that "projection." After all, this is the most fundamental, accurate way. You must realize, however, that any graphic representation of these data will be badly distorted, except near the equator where a degree of longitude covers approximately the same distance as a degree of latitude. Anywhere else, any image of the coverage is visually distorted, and the length of most lines is virtually meaningless, since the "length" is based on differences between latitude and longitude numbers. Such numbers do not provide a Cartesian two-dimensional space. (Recall the old riddle: where can you walk south one mile, east one mile, and north one mile, only to find yourself back at the starting point?* Not on any Cartesian x-y grid, for sure. Descartes, for his plane, insisted that a unit distance in the "x" direction be equivalent to a unit distance in the "y" direction.)

The Conversion Process: From GPS Pathfinder Office to ESRI Shapefiles

The Trimble Pathfinder Office software generates ESRI shapefiles simply, but, at the risk of being repetitive: all the caveats about being sure that datum, projection, coordinate system, and units are correct, apply.

A note on the shapefile data structure: ESRI refers to the representation of such geographic data as "a" shapefile. An ESRI shapefile is, in fact, a set of files which reside in the same folder. A shapefile named "abc," then, might consist of the files abc.shp, abc.shx, and abc.dbf. When you bring these files up in ArcMap you see only abc.shp in the table of contents. You should realize, however, that all the files beginning with "abc" are involved. Using the operating system to copy or rename them is a bad idea. Use the ArcCatalog capabilities (e.g., Manage Data Sources) to manipulate shapefiles.

After conversion you will find the files that constitute the shapefile in the Pathfinder Office Project Export folder.

The Files That Document the Export Process

The record of the conversion process is placed in two text files named something like exp1204a.txt (where 1204A indicates the month, day, and a sequence letter) and cl11315a.inf. These files will also be in the Export folder.

---

* At the North Pole. And at an infinite number of locations near the South Pole, such that you could walk south one mile to a circle around the pole that has a circumference of exactly one mile (or one-half mile, or one-third mile, . . .).

The "exp" file will look something like this:

```
Export Started...
Using Export Setup: AV_MDK#4
The following files in c:\gps2gis_ch6adata\river_av will be
exported:
     c111315a.cor

Reading file c111315a.cor
File c111315a.cor read successfully

1 input file(s) read.
83 position(s) read.
A total of 83 feature(s) read or created.
  83 point feature(s) created from GPS positions.
83 feature(s) exported.
3 output file(s) written to C:\Gps2gis_Ch6AData\River_AV\export
     posnpnt.shp
     posnpnt.shx
     posnpnt.dbf

The file c111315a.inf contains information on the settings used.

The file

c:\program files\common files\trimble\pfoffice\config\expfiles.txt

contains a list of the files created.
```

This file will give you a good idea of how the process went. This information is also available to you at the time the export process takes place; just ask to view the log of the export process. It is a good idea to at least look at the log when you do the export. Go back to the "exp" file if you run into any difficulty. (Note from the last lines that a third file is generated, buried deep in the general Pathfinder Office program files folder, that tells what files were generated—probably no new information there.)

The file that contains the GPS file name (in this case c111315a.inf) contains information about the setup used. This file is certainly worth reviewing if the data are to be used for any important purpose. Probably the file should be reviewed by at least two people if the data are to be published or used by the public or other people.

```
Setup Used:            AV_MDK#4
Export Format:         ArcView Shapefile
Data Type:             Features
Not In Feature Positions:One point per Not in Feature position
Export Notes:          No
Export Velocity Records: No
Export Sensor Records:  No
Export Menu Attribute As:Attribute Value
Generated Attributes:   GPS Time
```

```
                                 GPS Height
Position Filter Details:
Filter By:                       GPS Criteria
Maximum PDOP:                    Any
Min Number Of SVs:               3D (4 or more SVs)
Uncorrected:                     No
P(Y) Code:                       No
Differential:                    Yes
Realtime Differential:           No
Phase Corrected(Float):          No
Phase Corrected(Fixed):          No
RTK Float:                       No
RTK Fixed:                       No
Non GPS:                         No
Coordinate System:               Universal Transverse Mercator
Coordinate Zone:                 16 North
Datum:                           NAD 1927 (Conus)
Coordinate Units:                Meters
Altitude Units:                  Meters
Altitude Reference:              MSL
Distance Units:                  Meters
Area Units:                      Square Meters
Precision Units:                 Meters

Data Dictionary
---------------
Waypoint - Point Feature
    Label - String, Length = 100
    GPS Time - Time
    GPS Height - Numeric, DP = 3, Min = 0, Max = 0, Default = 0
PosnPnt - Point Feature
    GPS Time - Time
    GPS Height - Numeric, DP = 3, Min = 0, Max = 0, Default = 0
PosnLine - Line Feature
    GPS Time - Time
PosnArea - Area Feature
    GPS Time - Time
```

This information is descriptive of the process. Of course, your main interest is in the shapefile that is generated by the Export process. The Step-by-Step section that follows will lead you through the manufacture of a shapefile.

Bringing GPS Data to ArcGIS: Major Steps

A summary of the steps required to produce an ESRI shapefile from Trimble Navigation GPS data is as follows:

1. Collect data with a GPS receiver.
2. Load data into a PC—in the process SSF files are created.
3. Examine data graphically in the PC, and correct it, differentially and other- wise, as appropriate.

4. Convert data from Trimble format to an ESRI shapefile using Pathfinder Office (or other third-party software), being especially careful to use the proper parameters for datum, projection, coordinate system, and units.

5. In ArcCatalog examine each shapefile, and rename it (or copy it, giving it a new name). Repeat the process for as many shapefiles as you have.

6. Obtain other GIS data such as sets of data that you digitize, TIGER files, GRIDs, TINs, DEMs, DRGs, soils or image files, available commercially or from governmental sources.

7. Use ArcMap to visually integrate the converted GPS data with other GIS data.

STEP-BY-STEP

In this section there are projects you need Pathfinder Office to do (e.g., converting SSF files to shapefiles) but there are also projects that anyone with access to ArcGIS Desktop can do. The double line boxes are the Pathfinder Office projects; the single line boxes are the ArcGIS projects.

Pathfinder Office

Project 5A

In November of 1993 the students and faculty of the Department of Geography at the University of Kentucky participated in a cleanup of the Kentucky River. They took a GPS receiver on their trek; the antenna was mounted on the roof of a garbage scow (originally built as a houseboat). One file they collected, along the river from a marina to an "island" in the river, is C111315A.SSF. Using post-processing differential correction, it was converted to C111315A.COR. In this project you will make an ESRI point shapefile from the COR file, using the Trimble software and ArcCatalog.

Making a Shapefile from a COR File

{__1} *Start Pathfinder Office in the usual way.*

{__2} *In Pathfinder Office make a PROJECT named BOATTRIP_AV_yis (where yis represents your initials); set it up so that the associated folder is* ___:\GPS2GIS\RIVER_AV. *Switch to that project.*

{__3} *You should find a differentially corrected file C111315A.COR in your project's main folder. Open it in Pathfinder Office. Map it and Time Line it. The image you see should look something like a fish hook—in keeping with our nautical theme; the GPS track was taken over a period of about 45 minutes starting around 10:00 a.m.*

{__4} *Under Pathfinder Office Options set the Distance Units to "U.S. Survey Feet." Make the other distance variables relate to Feet as well. The North Reference should be True.*

{__5} *The Coordinate System should be selected by "Coordinate System and Zone." For the System use U.S. State Plane 1983. The Zone should be Kentucky North 1601. Measure altitude from mean sea level. Use the Defined Geoid. Make the Coordinate Units and Altitude Units U.S. Survey Feet.*

{__6} *Don't change anything in Style of Display* **unless** *the Coordinate Order is* **not** *North/East or the Scale Format is* **not** *1:X. The Time Zone should be Eastern Standard USA.*

The process of making an ESRI shapefile from a .COR file is complex enough, and error prone enough, that the Pathfinder Office software insists that you set the conversion up in advance. You do this with an Export Setup file, to which you give a name. You may use this file for later exports—either exactly as it is or somewhat modified. There are seven major facets to the Export Setup file. In no particular order, they are:

- Data
- Output
- Attributes
- Units
- Position Filter
- Coordinate System
- ESRI Shapefile

{__7} *Bring up the Export window, by either picking Export under the Utilities menu item or by clicking on the Export icon on the Utilities toolbar. (See Figure 5.2.) In the window, under Input Files, Browse to find C111315A.COR in ___:\GPS2GIS\RIVER_AV. Select it, then "Open" it.*

{__8} *The next step is to pick a folder that will hold the exported shapefile. If you have set up the project correctly the default folder shown should be:*

___:\GPS2GIS\RIVER_AV\EXPORT.

There are three sample export setups that work with ESRI products:

- Sample ArcInfo (NT & UNIX) Generate Setup
- Sample ArcInfo (PC) Generate Setup
- Sample ArcView Shapefile Setup

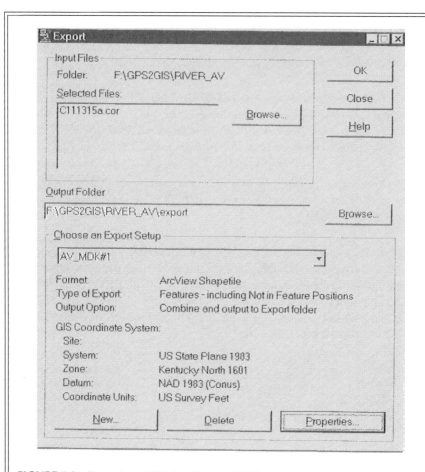

FIGURE 5.2   Exporting a GPS data file to a GIS format.

{__9} *From the drop-down menu under "Choose an Export Setup" find "Sample ArcView Shapefile Setup" and select it. Click "New" at the bottom of the window to bring up a "New Setup" window. (See Figure 5.3.) The idea here is to create a new setup that suits your needs, without changing the original setup. Choose "Copy of existing setup."*

{__10} *Replace the text in the "Setup Name" window with AV_yis#1 to rename the new setup. Then OK the New Setup Window.*

{__11} *A window should appear* entitled Export Setup Properties – AV_yis#1. The top portions should look like Figure 5.4. In this window are seven tabs, conforming to the list of seven facets listed above.*

---

* In the event that the setup already exists (perhaps because you have worked through this part of the text before), use the Properties button to bring up the window referred to here.

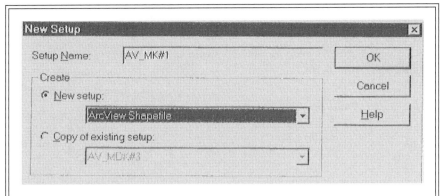

FIGURE 5.3   Creating a customized setup to make a shapefile.

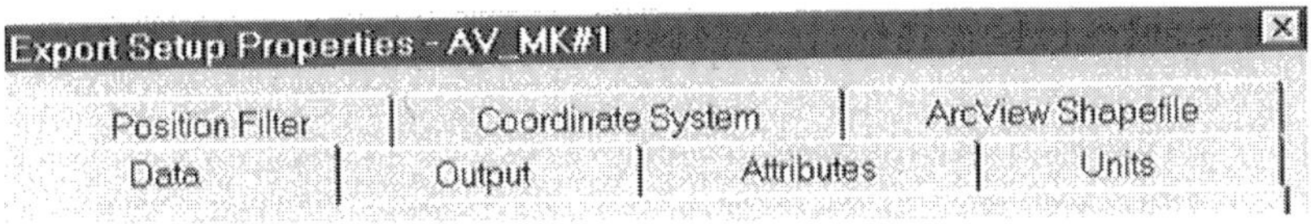

FIGURE 5.4   Set seven properties for a successful shapefile export.

{__12} *Start with the Data tab. You want to export "Features – Positions and Attributes." Pick Export All Features from the drop-down list. You want to include Not-in Feature Positions. And you want one point per Not-in-Feature position. See Figure 5.5*

{__13} *Under the "Output" tab, choose the option to combine all input files and place the results in the project export folder. For the System File Format choose DOS Files unless you are using some other system and the options are not grayed out.*

{__14} *Under the Attributes tab, let's export a couple of attributes as well as positions. Make checks by Time Recorded (under "All Feature Types") and Height (under "Point Features") as in Figure 5.6.*

{__15} *The Units tab: Here you will tell Pathfinder Office to use the Current Display settings, which should be U.S. Survey Feet, Square Feet, and Feet Per Second. Most of this window is "grayed out" because of other option choices you have made.*

{__16} *Move to the Position Filter tab. Make the Minimum Satellites "4 or more," for good 3D fixes. Accept any PDOP. Include only positions that are differentially corrected, i.e., "Postprocessed Code." The result should look something like Figure 5.7.*

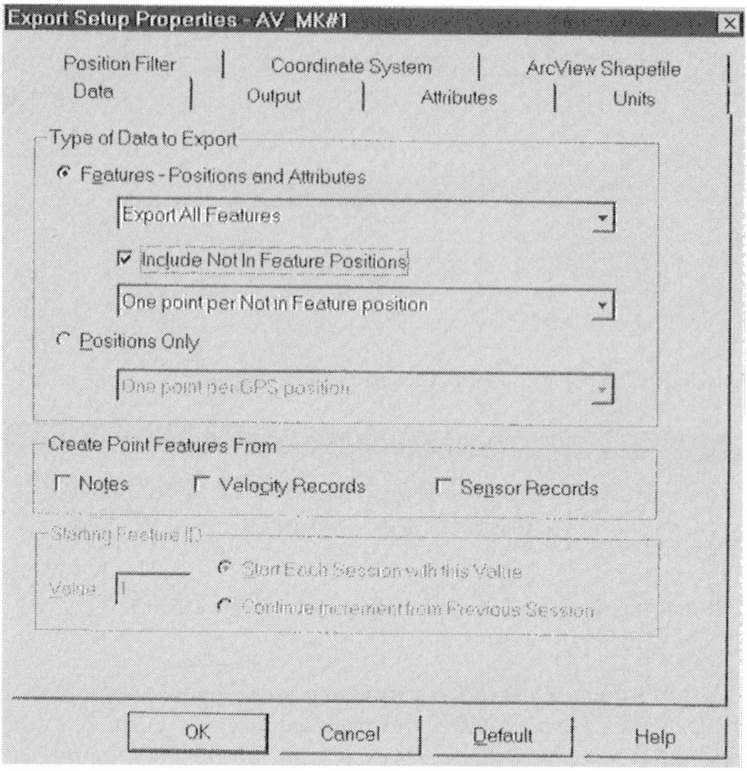

FIGURE 5.5   Defining the form of the data to be exported.

{__17} *Under Coordinate System, again use the current display coordinate system. Accept it after checking that it is U.S. State Plane, Kentucky North Zone 1601, NAD-83 (Conus), coordinate units and altitude units in Survey Feet, and the Altitude reference as MSL. Plan to export only 2D (XY) coordinates. If any of these display settings aren't right, back out of Export and correct the settings under the main Options menu. Make the new shapefile include an ESRI Projection file: Browse to Projected Coordinate Systems > State Plane > NAD 1983 (Feet) > NAD 1983 StatePlane Kentucky North FIPS 1601 (Feet).prj and click it to highlight it. Click Open. You will probably see a string of dots—it's okay.*

{__18} *Leave the options under "ArcView Shapefile" blank. Since you have now set all seven tabs, click OK to bring back the Export window.*

{__19} *Review the Export window. It should look about like Figure 5.2. If not, click "Properties" and fix things.*

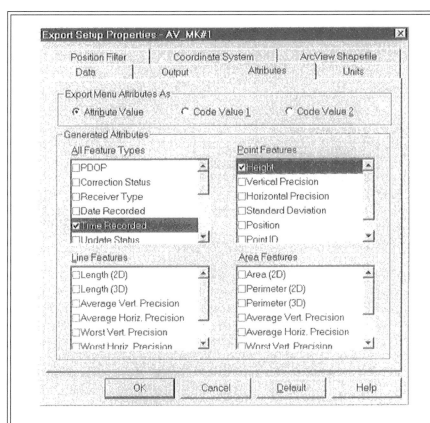

FIGURE 5.6   Defining the attributes to be exported.

{__20} *Click OK in the Export window. If you are asked about overwriting existing files, answer "Yes." You should then get an Export Completed window that looks like Figure 5.8. Click on More Details and read the resulting file. Dismiss that window and close its parent window.*

{__21} *If you now look at the project toolbar of the Pathfinder Office window, where the project you are working on is named, you will see the amount of storage space left on the disk that contains that project. You will also see a file folder icon. Click on it and examine the project folder: ___:\GPS2GIS\ RIVER_AV. Now pounce on the Export folder. You should see files that constitute an ESRI shapefile as well as the information files generated during the exporting process.*

C111315a.inf
exp1115a.txt (or some approximation thereof)
Posnpnt.dbf
Posnpnt.shp
Posnpnt.shx

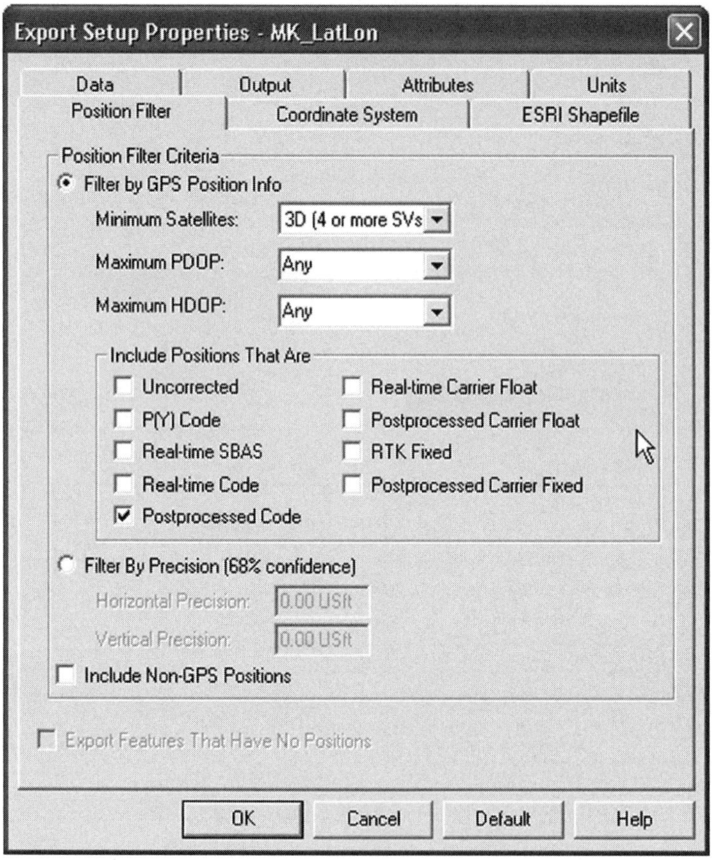

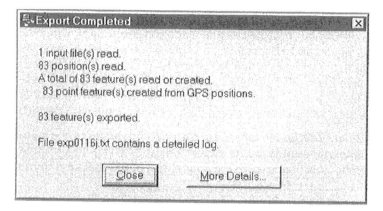

FIGURE 5.7   Defining the types of data to be exported.

FIGURE 5.8   Results of a successful export process.

If you don't see at least these files (perhaps with slightly different names), something went wrong and you will have to figure out what it was.

{__22} *If you pounce on a .txt or .inf file probably a text editor (e.g., WordPad or Notepad in Windows) will open showing the contents of the file. Examine the .inf and .txt files. Normally you don't need these files but they can be helpful if something goes wrong. When you are sure that everything is okay, dismiss the project folder window and minimize the Pathfinder Office window.*

See the Converted File in ArcMap

{__23} *Start up ArcMap. Make the window occupy the full extent of the screen. Under Tools, click Extensions. If your list of available extensions includes 3D Analyst and/or Spatial Analyst, put checks beside those names. Close the window.*

{__24} *Prepare to add data from ___:\GPS2GIS\RIVER_AV\EXPORT where you put the GPS track. You should see the name Posnpnt.shp in the left side of the Table of Contents window.*

{__25} *Add the data file Posnpnt.shp. You should see the same GPS track that you saw before in Pathfinder Office. This (see Figure 5.9) should tell you that you have turned a GPS COR file into an ESRI shapefile.*

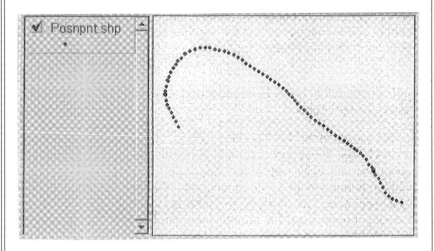

FIGURE 5.9   ArcGIS view of the Kentucky River GPS track.

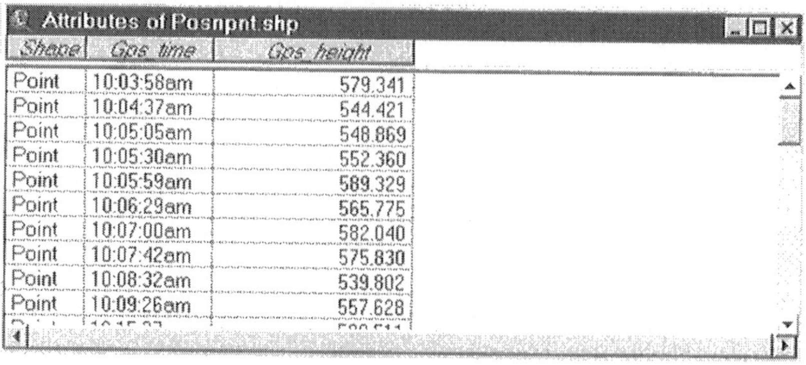
Attributes of Posnpnt.shp

Shape	Gps time	Gps height
Point	10:03:58am	579.341
Point	10:04:37am	544.421
Point	10:05:05am	548.869
Point	10:05:30am	552.360
Point	10:05:59am	589.329
Point	10:06:29am	565.775
Point	10:07:00am	582.040
Point	10:07:42am	575.830
Point	10:08:32am	539.802
Point	10:09:26am	557.628

FIGURE 5.10 Attribute table of the Kentucky River GPS track.

{__26} *Look at the attribute table: Attributes of Posnpnt.shp. The first few records of the 83 records are shown in Figure 5.10. You should see columns containing the time each fix was recorded and its height above mean sea level. Note the beginning and ending time for the data collection run. Note the wide variation of altitudes; since the river is very placid and almost level, this gives you an idea of what you can expect in the way of vertical accuracy for individual fixes—even those that have been differentially corrected. On the other hand, these data were taken a long time ago—things have improved somewhat. But the vertical will always be worse than the horizontal. Dismiss the table. Dismiss ArcMap.*

{__27} *Since Pathfinder Office restricts the names of shapefiles generated by its export utility to Posnpnt, Posnline, and Posnpoly, it is best to give the shapefile a name that says something about it. Start ArcCatalog. Navigate to __:\GPS2GIS\RIVER_AV\EXPORT\PosnPnt.shp and highlight it. Click the Contents tab at the top of the Data Frame to be sure this is the correct file. Click the Preview tab at the top of the Data Frame. You should see the GPS track.*

{__28} *Under the File menu click Rename. In the Catalog Tree, where PosnPnt.shp is boxed and highlighted type Boat_SP83_yis.shp. The name will change. To see the GPS track with other data, proceed to Project 5B below.*

ArcGIS

Project 5B: Looking at GPS and Other Data Using ArcGIS

In November of 1993 the students and faculty of the Department of Geography at the University of Kentucky participated in a cleanup of the Kentucky River. They took a GPS receiver on their trek; the antenna was mounted on the roof of a garbage scow (originally built as a houseboat). One file they collected, along the river from a marina to an "island" in the river, is C111315A.SSF. Using post-processing differential correction, it was converted to C111315A.COR. Using Pathfinder Office and ArcGIS this produced a shapefile called Boat_SP83 which is lodged, with some other data, in ___:\IGPSwArcGIS\River. In this project you will look at the GPS track along with other different types of data.

Look at the GPS Track in the Context of a Variety of GIS Data

The subject of this book is using GPS for input to GIS. Now that you have GPS data in ArcMap let's look at it with other GIS data. We have available several digital maps and images of a portion of the Kentucky River plus a couple of vector coverages of a few arcs from two USGS quadrangle maps. The county to the north is Fayette; that to the south is Madison. Of the two 7.5-minute quadrangles, the westernmost is COLETOWN; the other is FORD.

In the folder ___:\IGPSwArcGIS\RIVER exist the following:

- A line coverage* digitized from the Ford, Kentucky, quadrangle: FORD_VCTR
- A line coverage digitized from the Coletown, Kentucky, quadrangle: COLE_VCTR
- A TIGER line file (shapefile) of the streams of Fayette County, Kentucky: FAY_TIGER
- A TIGER line file (shapefile) of the streams of Madison County, Kentucky: MAD_TIGER
- A digital raster graphics file scanned from the Coletown topographic quadrangle: COLE_DRG.TIF
- A small orthophoto GeoTIFF image† (it is a digital orthophoto quadrangle that is 1/64th of a regular 7.5-minute USGS quad): COLE_DOQ64.TIF
- A Digital Elevation Model (DEM) in the form of an ESRI GRID consisting of "square pillars of elevation" that are approximately 30 meters square: COLE_DEM
- A vector line coverage of the elevation contour lines generated from the DEM: COLE_CONTOURS
- A Triangulated Irregular Network (TIN) showing elevations derived from the DEM, in the form of an ArcInfo TIN: COLE_TIN

---

* A coverage is the earliest form of ESRI vector data. It is being largely replaced by the geodatabase.

† If regular TIFF files are used to portray geographic areas a separate world file that provides the geographic coordinates of the TIFF must accompany them. GeoTIFFs contain the relevant world file; it is embedded. Trimble software does not recognize GeoTIFFs and hence requires the separate world file. ArcMap may use either setup. You may convert a GeoTIFF to a TIFF and a world file with the ArcInfo command CONVERTIMAGE.

- A vector (polygon) coverage of most of the Coletown quadrangle showing soil types: COLE_SOIL
- A vector polygon coverage of the Coletown geologic (surface rock)
  quadrangle: COLE_ROCK

In the steps below you will add these feature-based, grid-based, tin-based,
and image-based themes.

{__1} *Start ArcMap. Add as data the point shapefile Boat_sp83.shp from
IGPSwArcGIS\RIVER. This is equivalent to the shapefile that you may have
created in the preceding Project, but housed in a different folder.*

{__2} ***Add the feature based data sets from USGS 7.5 minute topographic quadrangles:*** *First, note that all these data sets are located in ___:\IGPSwArcGIS\
RIVER. The coverages FORD_VCTR and COLE_VCTR contain a few arcs
digitized (without great accuracy) from the topo quads Ford and Coletown; the
data have been converted to Kentucky State Plane coordinates in the NAD 1983
datum. In particular, an attempt was made to trace the banks of the Kentucky
River and some local highways from the quad sheets. Add as data the arc component of FORD_VCTR.* Use bright green (size 1) for the arcs of FORD_VCTR.
Add the arcs of COLE_VCTR making the symbol a bright red line.*

{__3} *Zoom to Full Extent. Observe. Then Zoom to the Layer Boat_SP83.*

You should see the GPS track, a few arcs outlining the river from the Coletown
topo sheet, and a few arcs from the Ford topo sheet†. The starting point of the
trip is in the southeast; at the other end of the track you can see the polygon
outlining the island.

{__4} *Measure the length of the trip. (_____ feet).*

{__5} *Change the viewing area and measure some other distances: Zoom up on
the arcs around the island. What is the length of the island (_____ feet), and
the width of the river near the island? (_____ feet).*

TIGER Line Files are a product of the U.S. Bureau of the Census. TIGER
stands for Topologically Integrated Geographic Encoding and Referencing.
TIGER files primarily contain street data—names, street numbers, census
tracks, zip codes, county and state codes and the geographic coordinates of these
features. But they also include other types of data that can be represented in

---

* To add the arc component of a coverage, double click on the coverage icon. Then pick "arc" and
    "Add."
† Yes, this short trip crossed the boundary between two quad sheets. Not only are you learning
    about GPS and GIS, you are also confirming the First Law of Geography: any area of interest, of
    almost any size, will require multiple map sheets to represent.

linear form, such as political boundaries, railroads, and streams. FAY_TIGER and MAD_TIGER represent the streams of their respective counties.

{__6} *Display FAY_TIGER.SHP as line data using bright blue as a color and "2" as the size. Notice the two streams that come into the Kentucky River along the GPS track. Notice also that the north side of the river is considered a Fayette County stream.*

{__7} *Display MAD_TIGER in the same fashion as FAY_TIGER but using a pale blue color.*

{__8} **Add as data the digital raster graphics (DRG) image of the Coletown quadrangle:** *Navigate to ___:\IGPSwArcGIS\RIVER and add COLE_DRG. TIF. (Decline the offer to build pyramids.) If necessary, pull its title in the table of contents (T/C) to the bottom and turn it on. (From now on, put point-based data at the top of the T/C, line-based data below them, and image-based or grid-based data at the bottom.) Observe. Zoom the view to the DRG layer to see what a USSGS DRG image looks like. Then zoom up on the area of the GPS track.*

{__9} *Look at the contour lines (on the DRG—zoom in as necessary) near the river. Obviously the elevation changes sharply, since the contour lines are close together.*

{__10} *Further zoom up on the bend in the river. What is the NORMAL POOL ELEVATION? _____*

{__11} **Display the orthophoto "tif" file:** *Navigate to ___:\IGPSwArcGIS\ RIVER and add COLE_DOQ64.TIF. Drag its title to next to the bottom of the T/C (with only DRG theme below it). Zoom so the entire orthophoto image, plus the island, is in the view window. You can get an interesting perspective on the images by flipping COLE_DOQ64.TIF off and on. Experiment with different levels of magnification.*

{__12} **Show a digital elevation model (DEM) that is in the form of an ArcInfo GRID:*** *From ___:\IGPSwArcGIS\RIVER and add data based on COLE_DEM. Drag its title to the bottom of the T/C. Turn off the two image themes above it; turn it on and make it the active theme. Right click the name*

---

* In order to see this GRID you will need the Spatial Analyst extension or the 3D Analyst extension. As indicated earlier, you may need to tell ArcMap to install the extension(s) in the current session. To do this, go to the File menu and click Extensions. Put a check mark by Spatial Analyst or 3D Analyst, or both.

and choose Properties > Symbology. Show Classified. Right click the color ramp and turn off the check beside Graphic View. From the drop-down menu pick Elevation #2. Under Classification choose 10 classes of elevation. Click "Classify" and choose "Natural Breaks." At around what elevation does most of the land exist? _____ feet. OK. Apply. OK.

{__13} *Notice the ranges of values of elevation. The lowest value in the area of the GPS track is, of course, on the surface of the river. Zoom in on the GPS track and use the identify tool to see if the river elevation, say between two GPS points, agrees with what you wrote before for the normal pool elevation. Again experiment with various levels of magnification, including full extent of the DEM.*

{__14} *Redisplay the DRG, but use the Identify tool to check out the map elevations based on the DEM. Click on the topographic lines map where the text shows elevations of 600, 700, 850, and 900 feet near the bend in the river. Note the agreement (or lack thereof) between what the topo map says and the values of the underlying grid.*

The contour lines on the DRG are simply those on the topographic quadrangle, which has been photographed. Another way of representing elevation on a GIS is through "isolines" or "contour" lines created from a DEM. Anywhere on a given line represents the same elevation. The contour lines in the COLE_CONTOUR line coverage were developed by a computer program that processed the DEM for that quadrangle.

{__15} *Display the arc component of the coverage COLE_CONTOUR, using a black line of width "1." Display this data set and the DRG only. With COLE_CONTOUR active, use the Identify tool to compare the elevation values of the black contour lines with those "printed" on the DRG. This process may work better if you zoom to a location away from the river, since the contour lines there are so dense that it is hard to see what is going on. You will notice that the contour lines from the two different data sets are close, but not congruent. Knowing the elevation at every point in an area is a tricky (actually impossible) task. These two data sets were derived in very different ways. They actually show a remarkable degree of agreement.*

TINs (Triangulated Irregular Networks), along with contour lines, and digital elevation models, constitute a third way of representing topography in GISs. A TIN represents elevation by tessellating* the plane with triangles. Each triangle

---

* A 50-cent word that means "completely covering an area with geometric figures without gaps or overlaps." Actually, it is the projection of the triangles on the plane that tessellate it, since the sum of the areas of the triangles is more than the area of the plane—since the triangles are not parallel to the plane—but you get the idea.

is positioned above the plane such that each of its three corners is at the proper elevation. The result, then, is a surface made up of flat triangular plates at different angles.

{__16} ***Make a layer from the Triangulated Irregular Network ArcInfo TIN:**** *Navigate to ___:\IGPSwArcGIS\RIVER and add data based on COLE_TIN. Drag its title to the bottom of the T/C. Turn off any image themes above it. Zoom in on the GPS track. Then zoom more to see the construction of the TIN. (If the surfaces of two or more triangles have the same slope you will not see the line separating them.) Again, with the DEM active, check out some elevations with the Identify tool.*

{__17} *Use the Identify tool with the TIN active. Optionally, edit the legend of the TIN. Clicking on a triangle's face shows the elevation of the cursor point, the slopes of the triangle, and the aspect (direction it is inclined toward). With ArcScene you could also view the TIN from positions other than straight above. You could spend a full day exploring the different ways to display TINs. Come back to this if you are interested. Get your guidance from the ArcMap Help files.*

{__18} ***Make a layer from an ArcInfo polygon soils coverage:*** *From ___:\ IGPSwArcGIS\RIVER, add data based on the polygon component of COLE_ SOIL. Drag its title to the bottom of the T/C. Turn off any image themes above it. Zoom to the layer. You will see arcs outlining the polygons but the polygons will be all the same color. Bring up the layer's properties. Click Symbology and pick Categories > Unique values. Pick "Minor1" from the Values Field. Use Basic Random for the Color Ramp. Punch Add All Values. Click Apply. OK. Dismiss the Legend Editor. You should see 70 different values represented in COLE_SOIL. In order not to have these all values cluttering up the T/C, click on the minus sign next to the data set name.*

{__19} *Zoom up on the part of the GPS track that lies within the soils theme. Using the Identify tool, determine the Minor1 classification of the soil type that predominates within the banks of the river. _____*

{__20} *Just to get an idea of the character of the soils coverage for the entire quadrangle zoom to the layer. The blank areas occur because soils data are not compiled primarily by quadrangle, but by county and quadrangle together.*

{__21} *COLE_ROCK shows the surface geology classifications. Display it as a polygon theme in much the same manner as the soils coverage. Display it with unique values for each "Name" of each rock type. As you have seen with the other GIS data sets, it gets pretty interesting in the area of the river.*

---

* In order to see this TIN you will need the 3D Analyst extension.

{__22} *Spend a few minutes looking at the various themes at different scales. One great advantage of a GIS is that you can make maps display what you want by turning various combinations of layers on and off. Play with that capability. When you are finished determine if your instructor wants you to save the Project so he or she can see it later. If so, follow the directions given as to where to save it and what to call it. Close ArcMap.*

Of course, there is a long story behind each of the types of data that we have looked at briefly here. If you are interested in a particular type or source of data I suggest that you look first at the Web for sources and then study Web pages or texts for more detail.

ArcGIS

Project 5C: ArcGIS Shows You Blunders Caused by Using the Wrong Datum

One of the central difficulties in using GPS for input to GIS is mismatched parameters, as I have stated so often that you are sick of reading it. Indeed, the issue of getting the parameters of geodatasets to match is a general problem with GIS. Let's look at what happens if we aren't careful, demonstrated by differences in the Universal Transverse Mercator (UTM) coordinate systems based on projections of the NAD27 (North American Datum of 1927) and WGS84 (World Geodetic System of 1984), which, by the way is virtually identical to NAD83 (North American Datum of 1983).

In this project we will use the now-preferred vector data structure for personal computers (PCs): file geodatabases. Here everything pertinent to the computer-based storage of geographic for a project is contained in a single relational database file.* (With shapefiles, you may recall or know, there are several files in the same folder. And coverages have a structure too obtuse to even mention here.)

A geodatabase contains geodatasets. Geodatasets have a number of characteristics and parameters associated with them, such as datum and units. A geodataset contains feature classes which all share the common parameters of the geodataset.

In this project we will look at a file geodatabase named Blunder_Prevention.gdb. It contains two geodatasets named: UTM_based_on_NAD27 and UTM_based_on_WGS_84. The data set UTM_based_on_NAD27 contains the feature class Boat_UTM_27. You can easily guess what the other geodataset contains.

Our plan is to look at both feature classes with ArcCatalog. We will observe that, though they look identical, they have somewhat different coordinates—just different enough to get us into trouble if we attempt to combine them.

---

* For an extensive discussion of geodatabases (and an almost complete understanding of ArcGIS desktop) see the author's textbook/workbook **Introducing Geographic Information Systems with ArcGIS**, Second edition, from John Wiley & Sons, ISBN 978-0-470-39817-3.

{__1} *Start ArcCatalog. In the Catalog Tree navigate to ___:\IGPSArcGIS\ River and expand it (with the little + sign next to it). Click on Blunder_ Prevention.gdb and read the status bar. It says _____. Expand it and navigate to UTM_based_on_NAD27. The status bar says _____. Expand it and navigate to Boat_UTM_27. Status Bar says _____.*

Click the Preview tab next to Contents. You see the all-too-familiar GPS track of the boat. The X and Y coordinates have been added to the attribute table. Use the Identify tool to click on the easternmost GPS fix. Write POINT_X and POINT_Y in the table below, to the nearest meter.

	Easting (X)	**Northing (Y)**
UTM NAD 1927		
UTM WGS 1984		
Difference		

{__2} *Now click on Boat_UTM_84 and write the coordinates. Calculate the differences between them. The northing difference is quite startling, especially since if you pick up a paper copy or a USGS quadrangle of the area it may say something like "to convert from NAD 1927 to NAD 1983 move the grid lines 4 meters south and 6 meters west." You have to understand that, although UTM is shown on the map, the grid lines described are the latitude and longitude grid lines, not the UTM coordinates.*

Pathfinder Office

Project 5D

{__1} *Back in Pathfinder Office, generate the shapefile that is a line rather than a sequence of points: Basically, you will repeat Project 5A, using the Export Setup AV_yis#1 except this time choose the "One line per group of Not in Feature Positions" option under the "Data" tab. Under the Attributes tab, select Length (2D) and Length (3D) as Line Feature attributes. When the export process is conducted for this Export Setup the GPS fixes will generate a line shapefile. Instead of PosnPnt.shp, you will get PosnLine.shp.*

{__2} *Display PosnLine.shp with ArcMap in a new view. Look at the attribute table. First observe that it is very dull (only one record) compared with the (hardly exciting) table that showed each point. The time given is the starting time of the file. Also look at the two columns in the file labeled Gps_length and*

Gps_3dlength. Gps_length is the sum of the segment distances, in feet, of the GPS arc, considering only the horizontal components (two-dimensional) of the coordinate points. Gps_3dlength is the sum of the lengths of the sequence of lines that connect the three-dimensional coordinate points. You would expect it to be longer, but not that much longer, unless the Kentucky River was a major whitewater river. The large difference is due, actually, to the GPS errors in recording the heights of the points that make up the arc.

Some points to consider: Suppose positions had been collected every 30 seconds on the boat trip, the positions were extremely accurate, and the boat stayed precisely in the middle of the river. Would the length of the two-dimensional arc, as found in the theme table, overstate or understate the length of the trip? _____. If there had been only small random errors in the GPS readings on this trip, what would be the effect of these on the length of the arc? _____. Do these two phenomena tend to reinforce or cancel each other? _____.

Project 5E

{__1} *In the ___:\GPS2GIS\RIVER_AV folder find a GPS file named KYCLAUTO.SSF. This is a track from the marina, north on an access road, north onto I-75 to Lexington's New Circle Road, and west on New Circle.*

{__2} *Convert this GPS SSF file into a shapefile compatible with FORD_VCTR and COLE_VCTR coverages.*

{__3} *Use ArcMap to display this shapefile with FORD_VCTR and COLE_VCTR.*

Project 5F

{__1} *Either use the data you took while transporting the GPS receiver and antenna in Project 2A or collect some new data. Find some GIS data that covers the same area in which you took the GPS data. If using TerraSync and Pathfinder Office, convert the GPS data with the appropriate parameters to a shapefile and view it with other GIS data. If using ArcPad, load the data onto a PC and view it with other GIS data.*

6 Attributes and Positions

*In which you learn how to collect attribute
information about the environment simultaneously
with collecting GPS position data.**

OVERVIEW

Overheard: Global Positioning System? A system for positioning the globe?

Obtaining GIS Attribute Data with GPS Equipment and Software

> **Juno ST and Pathfinder Office:** The Overview discussion that follows relates
> primarily to Trimble equipment and software. However it is general enough,
> and the ideas widely applicable enough, so that anyone starting out to collect
> attribute information with GPS position data could benefit from reading it.

> **ArcPad:** When creating a QuickProject you are also presented with a QuickForm.
> However, the QuickForm is not as complete as the Trimble Data Dictionary, so
> probably it is best not to use QuickProject when collecting complex attribute
> data. Use the feature capture capabilities with ArcPad, which are beyond the
> scope of this book.

From a software point of view, a GIS could be defined as the marriage of a graphic
(or geographic) database (a GDB) with other databases—most frequently a relational
database (RDB). These other databases—which contain **attribute** data about features in the GDB—are usually textual in nature, but sometimes consist of drawings,
images, or even sounds. (For example, you can key in, or click on, a street address
and be shown a photo of the house there.) The combination of a GDB and RDB
allows the user to make textual queries and get graphical responses (e.g., show with
a red X those streetlights which have not been serviced since August 1995) or, con-
versely, make a graphical query and get a response in text (e.g., indicate the daily
yield from parking meters in this area that I have outlined on this image of the city
using a mouse pointer).

---

* The projects in this chapter are somewhat advanced exercises. At times you may need to have and use
the manuals that come with the equipment and software. The instructions contained herein are more
general than those given previously.

If a GIS is a database with attribute information about geographical features, then it's more than reasonable to collect the attribute data at the same time the positional data are collected. Thus far in this text we have not done this with GPS. The process of adding attribute data occurred after we generated the GIS coverages, because our GPS files contained only points in 3D space. Suppose you collected a sequence of fixes along a two-lane road. Once the arc was depicted in the GIS, you would have to later add the "two-lane" fact to the record which related to the arc, if you wanted that information in the database.

Probably the most efficient and accurate way to use GPS to develop a GIS database is to collect the position data and the attribute data at the same time. Because a human operator is, usually, required to take the position data with a GPS receiver, it makes sense to have her or him enter the attribute data as well. Many GPS receivers, including those that we have discussed, allow this sort of data collection.

THE ORGANIZATION OF ATTRIBUTE DATA

The entry of attribute information into a GIS by using a GPS receiver is facilitated by a **data dictionary**, which is a hierarchical collection of textual terms stored in the GPS receiver's memory. With Trimble equipment the terms fall into three categories:

- Feature
- Attribute
- Value

Feature is used as it is in ArcGIS. It refers to the feature-type that is the subject of a shapefile or the feature class of a geodatabase. Examples of feature-types are parking meters, paved roads, and land use. A point (or the average of a set of points) collected by the receiver for a given feature may become the basis for a record in an attribute table. A sequence of points recorded along a line may also become the basis for a record attribute table.

Attribute is also a concept parallel to one in ArcGIS. Attributes are the "columns" or "fields" of a table. Suppose, for example, you are developing a GIS database about parking meters. In addition to the positional data collected at the meters automatically, you might wish to be able to record each meter's identifying number as one attribute, and the condition of the meter as a second attribute. Information about the monthly revenue could be added later as a third attribute.

Value refers to the actual numeric or character entries in the table. Continuing our parking meter example, you could enter a meter number as the identifier of a particular parking meter and select a menu item "fair" to indicate the condition of the unit.

As you can see, there is a hierarchy to these terms: a feature contains attributes; attributes are columns of values.

Once you have collected feature data with a GPS receiver, if you are converting these data to ESRI data sets, each feature-type* becomes a separate shapefile

---

* The word *feature* has at least two meanings. It can mean the subject of data set (e.g., historical markers) or it can mean a particular entity (e.g., historical marker number 876). So I will use *feature-type* to indicate a general class of features, and *feature* to mean a particular entity.

with its own feature attribute table. The table consists of the usual initial items plus columns for each attribute for which there are data. If a given feature (record) has a data value for a given attribute, that value becomes an entry in the feature attribute table.

THE DATA DICTIONARY

A data dictionary need not be long or complicated. For example, consider the following one, named Very_Simple. It contains only feature-types; it contains no attributes, so also no values.

Rocks (point)
Trees (point)

You could make this data dictionary on your PC using the Pathfinder Office software. Then you could transfer it to your Juno ST receiver.

To collect feature data with the Very_Simple data dictionary, you might begin a GPS file (for example, A010101A), setting up the receiver to record a feature data point every 10 seconds. You might then move around the area of interest. When you arrived at a tree, you would select the feature-type "Trees" from the Juno ST menu and collect a number of fixes at, say, 10-second intervals. You would then close the feature and, perhaps, move to the site of another tree, opening the feature, collecting data, and closing the feature. Should you encounter a rock you could open the feature-type "Rocks" and record fixes there. The fixes you record at each individual object are automatically averaged to produce a single point that approximates the position of the object.

Upon returning to your PC, you would use Pathfinder Office to transfer the file A010101A and generate shapefiles directly with the Pathfinder Office software. You would get three of them.

- POSNPTS.SHP
- ROCKS.SHP
- TREES.SHP

POSNPTS.SHP would consist of the points that the receiver recorded while you were walking between features.

ROCKS.SHP would contain a single point for each rock you visited. The point for a given rock would be the average of the fixes collected at that rock. That point would, therefore, provide approximate coordinates for that rock. A table would be formed with several records, one record for each rock. (The same idea would hold true for TREES.SHP.)

FEATURES WITH ATTRIBUTES ATTACHED

You might instead use a somewhat more complex data dictionary, Still_Simple, which might look something like this:

Rocks (point)
    Size (menu, no default value, required value)
        Small
        Medium
        Large
    Color (menu)
        Black
        Gray (default value)
        White
Trees (point)
    Size (numeric field, between 5 and 300, default value 10, two decimal places)
    Type (character field, maximum length 10)
Streets (line)
    Width (number of lanes, integer numeric field, min 1, max 10, default 2)
    Pavement (menu)
        Blacktop (default value)
        Concrete
        Other
        None
Intersections (point)
    T_intersection (menu, no default, required value)
        Yes
        No
Hydrants (point)
    Type (menu)
        Fire (default value)
        Lawn
StreetLights
    Bulb Type (menu)
        Incandescent
        Halogen
        Argon (default)
        Mercury

Assuming data were taken for all the feature types (Rocks, Trees, Streets, Intersections, Hydrants, and StreetLights) seven shapefiles coverages would result: POSNLINE.SHP, ROCKS.SHP, TREES.SHP, STREETS.SHP, and INTERSEC. SHP, HYDRANTS.SHP, and STREETLIGHTS.SHP. In addition to the standard items (columns, fields) in the shapefile table, ROCKS.SHP would contain the columns SIZE and COLOR. The values which could appear in these columns would be "Small," "Medium," and "Large" or no value at all for SIZE. What values could appear in the COLOR column? _____, _____, _____.

From the Environment, Through **GPS**, to **GIS**

The process of recording attribute data with a GPS receiver is a good bit more complex than simply recording position data, which is itself, as you know only too well, not a trivial matter. To record attribute data you have to go through several steps:

- Build a Data Dictionary with a computer. This can be done on a PC using the Pathfinder Office software.
- Load the data dictionary file into a GPS receiver (e.g., the Juno ST). This process is similar to transferring position files and almanacs from the receiver to the PC; the data sets simply go the other direction. The PC may contain a number of data dictionaries as may the Juno and most Trimble receivers.
- Take the receiver to the field and open a file. Select a particular feature-type (for example, a point feature) from the menu. While the unit is automatically collecting position information, you manually select the appropriate attribute and value items. When enough fixes have been collected, stop the data collection process for that feature (i.e., close the feature). If the feature-type is of the "point" type, the fixes obtained will be averaged so that a single point represents the given point feature.
- Continue to collect data in the field. The data collected when no feature is selected may ultimately become one shapefile. For each feature-type for which you collect data, a shapefile may be built. That shapefile will contain the number of individual features for which you recorded position information. Note that a given data collection session could result in a number of shapefiles, the data for which might all be contained within one GPS file. Linear features may be collected in a somewhat similar way.
- Close the GPS file. Open a new one if you like and collect additional feature data, or simply positional data, as you wish.
- Return to your office or lab and upload the files from your GPS receiver to the PC in the usual way.
- Differentially correct the positions in the files.
- Using Pathfinder Office, produce shapefiles.

To practice seeing and using the feature attribute collection abilities of GPS, do Projects 6A, 6B, 6C, and 6D.

STEP-BY-STEP

Attributes, **GPS**, and **GIS**

> This project requires Pathfinder Office if you want to go through it on a PC. However, it could be useful to read through it even if that software isn't available.

Project 6A: Demonstrating Feature Attribute Data

In March of 1999 on different occasions, a light plane was flown from Lexington to surrounding airports, recording GPS position and feature data. Usually a "touch and go" (brief landing and takeoff) was accomplished in which the aircraft wheels contacted the runway and a GPS fix (including the GPS altitude and local time) was recorded in the GPS receiver. Feature data, including the altitude reading from the aircraft altimeter and the published altitude of the airport, was likewise input. You will examine one of these flights: a trip from Bluegrass Airport near Lexington, Kentucky to the Montgomery County Airport near Mount Sterling, Kentucky. On this particular flight a full-stop landing was done at the destination airport.

(The data logging was done with a laptop computer, rather than a dedicated GPS receiver. The laptop system has several advantages over simply using a self-contained receiver-datalogger. For one, you have a full keyboard with which to enter data. Further, you have a screen with which you can display a map of the area and see your locations and progress in context. On the other hand, the laptop is less rugged and uses more power than a GPS receiver, and some laptop screens are difficult to read in bright sunlight. When you add the cost of the laptop, the total cost is more. But you may already have the laptop computer. As you can see, there are many considerations involved in what GPS datalogging system you should use for particular applications.)

{__1} *In Pathfinder Office set up the following parameters: Kentucky state plane coordinates, north zone (1601), NAD-83; survey feet for all linear measurements; time zone EST; Altitude MSL; North Reference Magnetic.*

{__2} *Open up the Map, Time Line, Position Properties, and Feature Properties windows. Make the Map window occupy most of the upper part of the left two-thirds of the screen and arrange the other windows as shown in Figure 6.1.*

{__3} *In the folder ___:\GPS2GIS\Flying_Tour, open the file R032720A.COR and the background geotiff files LEX.TIF and MTS.TIF. Represent Not In Feature data with medium red dots. Zoom to Extents.*

{__4} *Zoom up the westernmost image so that it fills the Map window. You should see most of Bluegrass Airport, including the northeast end (the departure end) of runway 4. (See Figure 6.2.) The GPS track was started just after takeoff. Using Position Properties, determine the altitude of the plane at the first GPS fix. _____. The published airport elevation is 979 feet.*

Airport runways are designated by taking their approximate magnetic heading (to the nearest 10 degrees) and dropping the final zero. So aircraft taking off or landing on runway 4 head approximately 40°, that is, almost directly northeast.

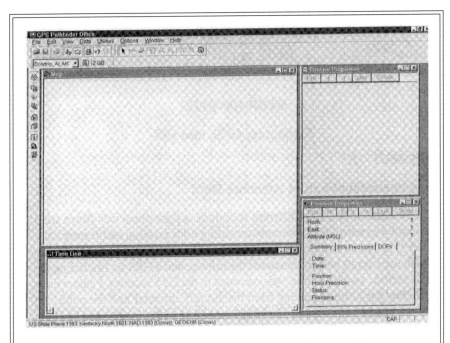

FIGURE 6.1   Screen layout of map, time line, position, and feature windows.

FIGURE 6.2   GPS track departing Bluegrass Airport to the northeast.

{__5} *Zoom up on the northeast end of the runway. You can see numbers painted on the concrete. Aircraft landing toward or taking off to the southwest would refer to this runway as "two-two"; 220° is the reciprocal of 40°. Use Zoom to Previous so that the entire airport image fills the map window.*

{__6} *In View ~ Layers ~ Features you will notice feature names of Altitude and Airport. Represent "Altitude" with a blue dot and "Airport" with a green dot. Pathfinder Office is aware that the Altitude and Airport layers exist because you have opened a file that contains those feature-types.*

{__7} *Make the Map window active by clicking on its title bar. Turn on Auto-pan to Selection. Using the position properties window, push "First" and then begin moving along the GPS track with the ">" button (or Alt-> key combination). Note the aircraft's altitudes during the climb out phase of the flight. At point 14 or 15, stop and examine the track. You should see a blue dot to the east of the selected fix.*

The blue dot represents a **quickmark**—a point feature that is made while moving along a path. A quickmark's position consists of an interpolation between the two fixes that bound it, in terms of time, one before and one after. Normally, as you will see later, a point feature's position is made up of the average of a number of fixes that are collected at the same spot. You can't do this while whizzing past a point of interest at 100 miles per hour, so some GPS equipment, such as the Aspen PCMCIA card with a laptop computer, allows you to collect quickmarks. A quickmark identifies a position and the operator can put in the attribute information later. (With other GPS equipment without the quickmark feature you simply have to take a point feature consisting of a single point at the right moment.)

{__8} *Continue moving along the GPS track. After fix 16 "Not in Feature: 16" will change to "Position: Quickmark." In the Feature Properties window you will see "MSL 0" and "Barometer 29.92."* These are default values for the point feature; that is, the operator triggered the quickmark that recorded the position but didn't enter any feature values.*

{__9} *Continue along the GPS track. Notice that the counting of Not in Feature fixes has restarted at 1. There are only three such fixes before the next quickmark point feature. Move ahead to that point feature. Here the operator recorded that the aircraft altitude (MSL) was 2,800 feet, according to the altimeter, which was set at 30.18—the value that the ground controllers had transmitted to the pilot before takeoff. What was the GPS altitude? _____*

---

* Aircraft altimeters must be set to the local sea level barometric pressure if they are to accurately represent altitude. The value 29.92 was used as the default because it is "standard pressure." That is, at sea level, on an average day in an average place, the atmospheric pressure will support a column of mercury 29.92 inches in height.

{__10} *Continue along the GPS track with the ">" button. You may note that there are some 60-plus fixes remaining and that the scenery has become very dull. We can speed things up. Make the Feature Properties window active by clicking on its title bar. If you pause the pointer over the ">" button* **in that window** *you will see that it will take you to the next feature. Click that button.*

{__11} *You should see a DOQ of the airport near Mount Sterling. Just as the plane landed a quickmark was taken. Examine the information in the Feature Properties window.*

The Airport Name was selected from a menu in the data dictionary (we knew which airports we would be flying to; IOB is the airport designation). The altitude from the altimeter was read directly from the instrument and typed into the laptop computer. The time was local time, based on UTC time taken automatically from the receiver. The published altitude, the runway number,* and the runway length were input after consulting an aeronautical chart. Notice the differences between the altimeter altitude (1,040 feet), the published altitude (1,020 feet), and the GPS altitude (taken from the Position Properties window) of 1,012.644 feet. The differences in these altitudes indicate why one doesn't want to use an altimeter altitude or a GPS altitude (even a corrected one) as the sole determiner of altitude when getting an aircraft on the ground. A matter of 4 or 5 feet vertically can make the difference between a smooth landing and a controlled crash.

{__12} *You can experiment by clicking on various icons in the Time Line window, in the Feature Properties window, in the Position Properties window, and on the Map window. Sometimes these windows don't seem to be completely coordinated, but you can always ultimately get to view a given entity in several ways. Practice stepping through the various features from the first to the last.*

{__13} *In the Utilities menu (or on the Utility toolbar) select Data Dictionary Editor ~ File ~ Open. Navigate to*

___*:\GPS2GIS\Flying_Tour\Apt_Info.ddf*

and open it. This is the data dictionary file (ddf) that was used to create the features in R032720A.COR. Click on the various Features and Attributes to better understand the relationships between features, attributes, and values. I'll say much more about data dictionaries and their use in the following projects, but this exercise should give you the general idea of how it all fits together.

---

* If you zoom up on the southwest end of the runway you can see the runway number. Then pan along the path the aircraft took. Follow it into the parking ramp area. Then Zoom to previous.

Three Linked Projects 6B, 6C, and 6D: Utilizing Feature Attribute Data

The following three projects (6B, 6C, 6D) relate to obtaining and converting feature attribute information as well as position data. If you have a Juno receiver and Pathfinder Office software, and you want to collect data in the field, do all three projects.

If you do not want to take time to collect data, but do want to see most of the conversion process, do Projects 6C and 6D.

If you only want to see the results of feature data collection, and you have ArcGIS available, do Project 6D. In any event, you should plan to read through all three projects.

{__1} *To do any of the projects, you should make a separate folder to hold the data. Assuming you have done previous projects and have a MyGPS_yis\DATA_yis folder, make a new folder named FEATURES under it. Create a new Pathfinder Office Project, named FEATURES_yis, that references the new folder:*

___:\MYGPS_YIS\DATA_yis\FEATURES

You will need this folder if you do any of the "attribute" projects. Make FEATURES_ yis the current project.

This project requires Pathfinder Office and a Juno (or other Trimble) receiver.

Project 6B: Obtaining GIS Attribute Information with GPS Equipment

In the steps below you will make, in ___:\MYGPS_YIS\DATA_yis\FEATURES, a data dictionary using Pathfinder Office and the Still Simple Data Dictionary:

{__2} *In the Utilities menu select Data Dictionary Editor. Type Still_Simple as the name. Type in any Comment you wish. Plan to use the form, if not the exact text, from the Overview as a model for your data dictionary. (If you are in an urban area where rocks or trees are scarce, perhaps you could substitute TELE_POLES and PARK_METERS.)*

A data dictionary is built with a series of windows. You are prompted for feature-type name, attribute, and other information, as appropriate. (Ignore fields that ask for user codes.) "OK" enters the particular entity; "Done" records your choices so far, then takes you back a level in the hierarchy. A "default value" is the one selected by the Juno ST at data collection time if you don't select another. As you build the dictionary you will see it appear in a box on the right of the screen. You should be able to make "Still_Simple" with just the information found in the Overview section and the steps below.

{__3} *Click New Feature. The first Feature Name will be Rocks. You may add a comment here if you wish. The Feature Classification is Point. Under Default Settings leave the Logging interval at 5 seconds and the Minimum Positions at 1. The Accuracy must be Code, not Carrier. Okay the window. Notice that, back in the Data Dictionary Editor window, a record is being made of your progress, under Features.*

{__4} Click New Attribute. In the New Attribute Type window "Add" a "Menu" attribute. The name of the Attribute would be "Size." To specify the possible menu values click New. Type "Small" to make an Attribute Value and click OK. Now add "Medium" as an Attribute Value. Then "Large." Close the "New Attribute Value—Menu Item" window. Okay the "New Menu Attribute" window. Close the "New Attribute Type" window.

{__5} From what you have just learned construct the Color attribute of Rocks with the menu values Black, Gray, and White. Examine the specifications of the Still_Simple Data Dictionary described earlier in this section for the details.

*{__6} **Observe and experiment with the Data Dictionary Editor window:** Select Rocks. Select Color. Use the F8 key to Edit the spelling of "Color" to make "Colour" and add "Pink" to the list of Menu Attribute Values.*

{__7} Delete "Pink" as a possible value. Go back to the American spelling of "Color."

{__8} With the F3 key add the New Feature "Trees" with "Size" as a Numeric attribute with two decimal places. The Minimum should be 5; the Maximum 300. The default value—one that will be used if the person doing the datalogging fails to put in a value—should be 10.

{__9} Complete the data dictionary using Still_Simple as a guide. Figure 6.3 gives an idea of what it should look like, although there are many more facets to it than can be displayed in a single image.

{__10} Once your dictionary is complete, step through the features and attributes (by clicking on them) while examining the pane that gives information about the possible values. Use the edit capabilities of the Data Dictionary Editor to make any needed corrections.

It is important to get the data dictionary correct NOW. Once you get into the field you are pretty much stuck with what you have done. And once you have collected data with a particular data dictionary you become even more wedded to it. Get it right to begin with!

{__11} Under File ~ Save As, allow the suggested "data dictionary file" (ddf) name Still_Simple.ddf to stand, but put the file in ___:\GPS2GIS\DATA_yis\ FEATURES. Dismiss the Data Dictionary Editor.

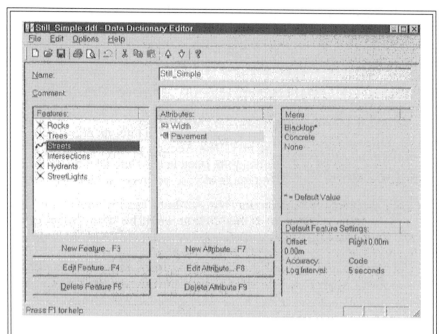

FIGURE 6.3   Making a Data Dictionary.

{__12} *Connect the Juno receiver to the PC as usual, setting the proper communication parameters in both Pathfinder Office and the receiver. Select "Data Transfer" from the main menu of the receiver.*

{__13} *In Pathfinder Office, Transfer the almanac to the PC, just to be sure the connection is working.*

{__14} ***Transfer the "Still_Simple" data dictionary to the Juno:*** *Press the Send tab (this time the computer is sending information to the receiver, rather than the other way around) and then Add ~ Data Dictionary. When the Open window appears navigate to ___:\GPS2GIS\DATA_yis\FEATURES. The file name Still_Simple.ddf should appear in the window. Click on it, and then click "Open." "Transfer All" should send the file on its way. When done, check on more details, then dismiss the now unneeded windows.*

{__15} *In the Juno, under Data > File Manager tap Dictionaries under Choose File Type. Click on Still_Simple. Under Options tap Edit dictionary. Examine the data dictionary you just loaded. Then click Cancel and abandon changes.*

{__16} *Disconnect the Juno from the computer. Turn it off.*

{__17} *Take the Juno into the field to begin collecting position and attribute data, according to the Still_Simple data dictionary. Turn it on. Start TerraSync.*

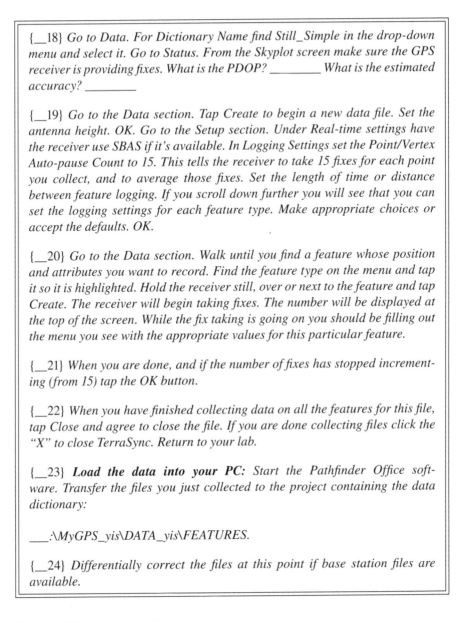

{__18} *Go to Data. For Dictionary Name find Still_Simple in the drop-down menu and select it. Go to Status. From the Skyplot screen make sure the GPS receiver is providing fixes. What is the PDOP?* _____ *What is the estimated accuracy?* _____

{__19} *Go to the Data section. Tap Create to begin a new data file. Set the antenna height. OK. Go to the Setup section. Under Real-time settings have the receiver use SBAS if it's available. In Logging Settings set the Point/Vertex Auto-pause Count to 15. This tells the receiver to take 15 fixes for each point you collect, and to average those fixes. Set the length of time or distance between feature logging. If you scroll down further you will see that you can set the logging settings for each feature type. Make appropriate choices or accept the defaults. OK.*

{__20} *Go to the Data section. Walk until you find a feature whose position and attributes you want to record. Find the feature type on the menu and tap it so it is highlighted. Hold the receiver still, over or next to the feature and tap Create. The receiver will begin taking fixes. The number will be displayed at the top of the screen. While the fix taking is going on you should be filling out the menu you see with the appropriate values for this particular feature.*

{__21} *When you are done, and if the number of fixes has stopped incrementing (from 15) tap the OK button.*

{__22} *When you have finished collecting data on all the features for this file, tap Close and agree to close the file. If you are done collecting files click the "X" to close TerraSync. Return to your lab.*

{__23} **Load the data into your PC:** *Start the Pathfinder Office software. Transfer the files you just collected to the project containing the data dictionary:*

___*:\MyGPS_yis\DATA_yis\FEATURES.*

{__24} *Differentially correct the files at this point if base station files are available.*

PROJECT 6C

TerraSync: GPS2GIS\FEATDEMO

I recommend that, even if you collected your own data in PROJECT 6B, you work through these PROJECT 6C exercises with the supplied data; then use the same process with your own data. (These projects deal with lines, and points within these lines, which I have not told you how to do. But it's not hard, if you want to do some exploration and extrapolation on your own.)

{__1} *Start Pathfinder Office. Make (or change to) a Project named Feature_ Demo. It should point to*

___*:\GPS2GIS\Feature_Demo.*

{__2} *Arrange the Map, Time Line, Feature Properties, and Position Properties as you did in PROJECT 6A (See Figure 6.1.)*

{__3} *Set up the following parameters: Kentucky state plane coordinates, north zone, NAD-83, survey feet for all linear measurements, time zone EST.*

{__4} *Open F121723C.COR and, as a background file, FIVE.TIF. With View > Layers > Features make:*

- Not In Feature data a thin red line
- Streets a thin yellow line
- Rocks a black dot with a yellow background
- Trees a green tree
- Intersections a black cross with a yellow background
- Hydrants a red hydrant
- Streetlights a gray streetlight

{__5} *Experiment with the Time Line (you can separate the features by zooming up in that window with the magnifying glass). With Auto-pan to Selection active, experiment with moving along the track with the Position Properties window. And then do the same with the Feature Properties window. Since the DOQ is not very high resolution about all you can verify from the image is that the feature Intersection does indeed show up where there are intersections. Trees and such are not distinguishable. See Figure 6.4.*

Make Shapefiles from the Feature-types

The process of making ESRI shapefiles from GPS files containing feature attribute data is virtually identical to that described in Chapter 6. The primary difference is that, instead of getting just one shapefile (posnpnt.shp or posnline. shp) you will get several:

- Not In Feature data will be in POSNLINE.SHP.
- Streets will be in STREETS.SHP.
- Rocks will not be represented as a shapefile since we didn't collect data for this feature.
- Trees will be in TREES.SHP.

- Intersections will be in INTERSEC.SHP, since Pathfinder Office only outputs the first eight characters of a data dictionary feature-type as an ESRI shapefile name.
- Hydrants will be in HYDRANTS.SHP.
- Streetlights will be in STREETLI.SHP.

FIGURE 6.4   Using Pathfinder Office to view collected features.

{__6} *Start with the Export utility. For the input file use F121723C.COR, which is in ___:\GPS2GIS\Feature_Demo (pointed to by the Project named Feature_Demo).*

{__7} *The Output Folder should automatically come up as*

___:\GPS2GIS\Feature_Demo\EXPORT. Fix it if it doesn't.

{__8} ***For practice make a New Export Setup:*** *Press the button "New Setup." From the drop-down list select ESRI Shapefile. The Setup Name will appear as "New ESRI Shapefile." Change it to "AV#4_yis." A window, Export Properties, will appear.*

Under the Data tab: For "Types of Data to Export" pick "Export All Features" from the drop-down list.

For Output: Combine all input files and output them to the project export folder.

Under Attributes: Output the Time Recorded for All Feature Types. For Point Features use the Height and the Point ID. For Line Features select the 2D Length, the 3D Length, and the Line ID.

Units: You can use the current display units, since those let you see both the GPS track and features as well as the DOQ image. (You would have to be careful using the setup AV#4_yis in the future, since it would use the display units in effect at that time.)

Position Filter: Make sure Differential is checked under "Filter by GPS Position Info."

Coordinate System: Again, you may use the current display coordinate system.

The ESRI Shapefile tab, which allows you to Export Tracking Themes, needs no attention here.

ESRI has produced an ArcGIS extension called the Tracking Analyst that allows real-time tracking of GPS receivers, or "playing back" data taken by such receivers. Basically, the Tracking Analyst allows the user to plot GPS data as it is being acquired, by virtue of radio transmissions, perhaps by communication satellite or cellular telephone, of the GPS data to a location that then relays these data to the ESRI software. If, for example, a bus company wanted to know the whereabouts of each bus and whether it was on schedule, it might use the Tracking Analyst to display this information in real time.

Click OK to exit the Export Setup Properties window. If you need to make a change or to be sure you've made the right choices, you can select "Properties" in the Export window.

{__9} When you are satisfied with the Export window, click OK. Pathfinder Office will commence making the shapefiles.

{__10} Click the folder icon on the Project toolbar to bring up the project folder. Double-click the Export folder icon. You will see three files for each feature-type. For example, HYDRANTS.SHP, HYDRANTS.DBF, and HYDRANTS. SHX. HYDRANTS.DBF is a database file that contains the attribute information about the hydrants. This information will appear in the ArcGIS table of HYDRANTS.SHP. HYDRANTS.SHX is a file that ArcGIS uses for its own purposes.

Now you may

(a) examine your own GPS data file and turn it into ESRI shapefiles (use your Project named FEATURES_yis, that references the folder:

___:\GPS2GIS\DATA_yis\FEATURES)

OR*

(b) proceed to PROJECT 6D to examine the shapefiles you just made in ArcGIS.

---

* This OR is called, in logic, the inclusive OR. It means either (a) or (b) or both. The exclusive OR means (a) or (b) but not both (a) and (b). English has both the inclusive OR and the exclusive OR; you must decide which is meant, based on context. Inclusive: "Would you care for cream OR sugar with your tea?" Exclusive: "You have won the contest and you may have a trip to Hawaii OR a new TV set."

PROJECT 6D: VIEWING SHAPEFILES

Here's a summary, mainly for those who didn't do the previous two projects: In PROJECT 6B we completed a data dictionary, installed it in the Juno, collected data under the structure of that dictionary, and brought those data back to the PC. In PROJECT 6C we looked at some similar sample data, and generated ESRI ArcGIS shapefiles.

From this point forward in the text, anyone with access to ArcGIS may process the files that resulted from the previous work.

{__1} *Start ArcMap. Make its window occupy the entire screen.*

{__2} *Add the data ___:\IGPSwArcGIS\Feature_Demo\EXPORT\STREETS.SHP.*

{__3} *Open the table for STREETS.SHP. For the streets feature that was collected you should see a record giving both the features you specified when the data were taken (width and pavement) plus the features that were specified when you generated, with the export setup, the shapefiles. Any numeric units will, of course, be those you indicated at the time of export. Dismiss the table.*

{__4} *Add the Image Data Source FIVE.TIF from ___:\GPS2GIS\Feature_Demo as data.*

{__5} *Add the feature-based theme INTERSEC.SHP. Make the cursor active with the Identify icon. Click on each intersection and verify that the T-Intersection designation (Yes or No) is correct. See Figure 6.5.*

{__6} *Open the other shapefiles generated in the preceding project. Look at their attribute tables. Make each streetlight a bright yellow.*

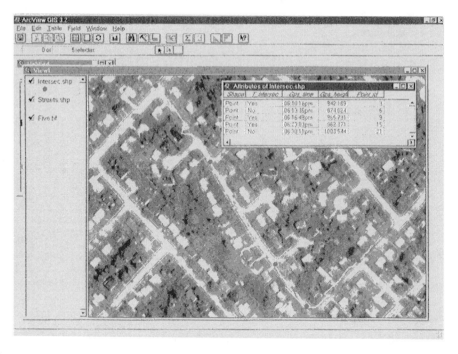

FIGURE 6.5   After Export, using ArcGIS to view features.

{__7} ***Display the "Bulb_type" for the streetlights on the map:*** *With the Identify tool (set it up for street lights) click each streetlight feature.*

Now Use Your Data

{__8} *If you collected data in PROJECT 6B, go through the procedures of PROJECTs 6C and 6D, using your own data, using appropriate datum, coordinate system, projection, and units. Use the folder ___:\MyGPS_yis\DATA_yis\FEATURES that is specified by the Project you made in PROJECT 6B.*

7 GPS Mission Planning

*In which you learn how to plan for
important GPS missions.*

OVERVIEW

PLANNING A GPS DATA COLLECTION SESSION

Fact: In a given area, the GPS data collection process works better at some times than at others. Determining a good (or just satisfactory) time to take data is called **"mission planning."**

Several years ago, this discussion of mission planning would have had to appear at the front of this book, because there were so few GPS satellites in orbit that you had to go into the field at specified times to collect data. Now with the NAVSTAR system at full operational capability (31 satellites up and healthy in 2008), you can almost always "see" enough satellites to get a pretty good fix. But is it a fix that is accurate enough to meet your needs? (Our needs in this book were simply to show you how the system worked. So we left the mission planning details until now.)

It might have been simpler if the GPS satellites were simply parked over various areas of the world so that, in a given location, you always dealt with the same set of satellites in the same positions. Unfortunately, this would mean that they were always parked directly over the equator, because, due to the laws of physics, that's where all geostationary satellites are. (If they were orbiting the Earth anywhere else, as they would have to be in order to provide worldwide coverage, they would be moving with respect to the ground underneath.)

So the rule for GPS satellites is "constant change." The number of SVs your receiver can track changes. The geometric pattern they make in the sky changes. The local environment in which you want to take data changes—perhaps there are objects that are blocking signals from reaching your antenna. If you want to collect data at the optimum times—for examples, when the DOP is very low, when there are enough satellites in the right part of the sky, or when there are a large number of satellites available for overdetermined position finding—you might want to use mission planning software.

Planning software will also help you if your data collection effort might be impeded because there are barriers between you and some part of the sky. You may use the mission planning software to simulate such barriers and to tell you when the satellite configuration will be such that you can take good data anyway.

ALMANACS

A GPS almanac is a file that gives the approximate location of each satellite. Actual position finding does not depend on the almanac; a much more precise description of each satellite's position is needed for a GPS fix calculation. (Precise descriptions of the orbit of a given satellite come from an ephemeris message—a broadcast that may change hourly—by the satellite itself.) Rather, the almanac is part of the general message which comes from **each** satellite, describing the orbits of **all** satellites. The almanac gives parameters from which the approximate position of all satellites can be calculated at any time t in the future.

Because it doesn't have to be very precise, an almanac may be good for months. Left to themselves, GPS satellites deviate little from their projected orbits. However, the DoD may move a satellite forward or back within its orbit—perhaps to replace a failing satellite or to make room for the launch of a new one. Activities like this make it necessary to collect almanac information frequently. In fact, the Juno ST or the Magellan MobileMapper running ArcPad collects an almanac every time it is outside and tracking one or more satellites for a minimum of about 15 minutes.

This almanac can be transferred to your PC. Although the receiver will not tell you the almanac's age, planning software will tell you if an almanac is more than a month or so old. An ephemeris file, from which planning software may derive an almanac, may also be loaded onto a PC across the Internet from the Trimble Navigation Web site.

Almanacs are used for at least two purposes:

* giving information to the receiver so that it can locate the satellites more quickly, and
* providing information to PC software so it can tell the user the best times to collect data.

It is this second purpose we examine here: using planning software to tell us when we can get the best data at a given location.

USING MISSION PLANNING SOFTWARE

Trimble Navigation provides software to help the GPS user determine the best time to collect data: a program running under Microsoft Windows. Basically, the software gives you information about the expected behavior of the satellites. This is information you need for planning important GPS data collection sessions. To see an illustration using the mission-planning software, do Project 7A. To see the effects of poor mission planning, do Project 7B.

STEP-BY-STEP

PLANNING A GPS DATA COLLECTION SESSION

You will learn about mission planning in the next three projects. Project 7A will use an old almanac with a program called Quick Plan which was provided in earlier versions of Pathfinder Office. It will illustrate the principles behind mission planning. In Project 7B you will see the results of poor planning (resulting in missing data), again by looking at the past. In Project 7C you will download, for free, the newest Trimble planning software, which allows use of not only NAVSTAR GPS satellites but also Russian and European satellites, if you have a receiver that can receive their signals. You will use this software to look at the present with a recent almanac. Since by that time you will know the principles of mission planning, you will figure out how to use the new software from the documentation and the help files. Should you wish to go immediately to the new software you may download it from http://www.trimble.com/planningsoftware_ts.asp.

Project 7A

{__1} *Start Pathfinder Office. Make a Pathfinder Office project called Planning_yis. Have it reference the folder Planning in the ___:\GPS2GIS directory.*

{__2} *With the operating system find a file named Plan.exe. Pounce on it.*

{__3} *When Quick Plan starts (click the title bar to make it occupy the full monitor screen) you will be asked to select a date for which you want to plan. As this program will only be used for demonstration purposes, we'll use a date from the past: September 15, 2006. So in the field in the lower right corner of the Select Date window type 09/15/2006 and OK the window. Skip the warning about the date being old.*

{__4} *In the Edit Point window that appears you can inform Quick Plan, by a variety of methods, the location at which you plan to take data. Let's just experiment with the three ways you can choose a location. Click on World Map. Either a Mercator projection or an image of a globe will be displayed. Let's get to the globe, which will look something like Figure 7.1. If you are not there already, click on the map a time or two until the "Globe" option appears. Click it. Click the globe. A dot cursor will jump to the closest city that is in the database. Its longitude and latitude will be displayed along with the city name. As you move the dot cursor the names of the cities it lands on will be displayed.*

{__5} *Find the city in which you would most like to take GPS data (or just visit). You may have to turn the globe left or right, up or down. Neat, huh? Turn off the Rivers. Enhance your knowledge of geography. Click the "About..." button.*

{__6} *With "Take city" checked, click on Mexico City. What are its geographic coordinates? _____, _____. Click OK to get back to the Edit Point screen.*

{__7} *Pick "World Map" again and then get a Mercator projection of the Earth. Click on it until you can move the pointer around the screen so that city names*

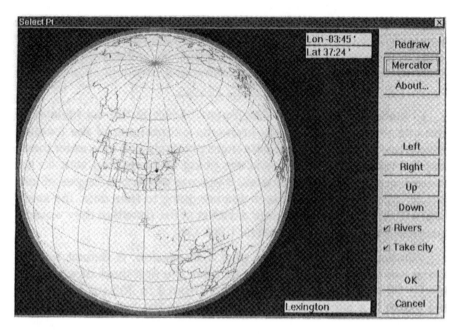

FIGURE 7.1   One way to select a city in Quick Plan.

appear at the bottom and the graticule coordinates show up at the top. After click-ing on a Hawaiian city, magnify the Hawaiian Island chain. Again. Again. Again. Demagnify and Magnify as necessary so you can click on Kailua Kona on the island of Hawaii. You may recall this city from Chapter 3.

{__8} *Redraw the map so it goes to its extents. Select Miami, Florida. Click OK. If "Take City" is checked and you click on a city and OK the window, the coordinates of that city will appear in the Edit Point window. Click OK. Write the name and coordinates of Miami: _____,_____.*

{__9} *Finally click again on "Keyboard." Look at the coordinates of the city you selected. If you knew the coordinates of the location where you were going to take data more precisely you could enter them, but the planning elements you are going to be dealing with do not require very precise coordinates. If you are within one degree of longitude and latitude of the place you are going to take data, the satellites will rise and set within 4 minutes of the time that they would if you had the precise coordinates. Click OK. OK the Edit Point window.*

{__10} *Under the Session menu, pick Edit session. A default session name will appear as a four-digit number, the first three of which will be the Julian date, num-bering the days from 1 to 365 (or 366), while the fourth is a sequence digit. Click Edit. In the "all points" window you should see the point corresponding to the city*

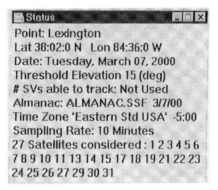

FIGURE 7.2   A Quick Plan status window based on an almanac.

you selected last. Click on it—then click "<Add<" to add the point to the empty list on the left side of the window. Click OK.

{__11} *A Status window will appear. (If the window such as that in the next figure (Figure 7.2) is not on the screen, select "Options" from the main menu, followed by clicking on "Show Status.") Examine the status window. Check its text to be sure everything is correct. It may not indicate what satellites are tracked nor what almanac is used (as the one in the figure does). We'll fix that.*

{__12} *From the Options menu pick "Time Zone." From the zone name drop-down menu, find Eastern Day USA and select it. Click OK.*

{__13} *Under the Options menu pick Almanac. In the "Load File with Almanac" window navigate to C:\GPS2GIS\Planning\ALMANA~1.SSF. Select it so that it appears under File Name. Click OK. The Status box will be much improved. Check that the correct almanac is employed. Notice the number of satellites considered in the planning session; they are enumerated. For an example see Figure 7.2. Dismiss the Status window (you can always get it back: Options ~ Show Status).*

{__14} *Select "Graphs." Pick "Number SVs and PDOP." A graph with two parts will appear. Make it full screen. The graph should now appear something like Figure 7.3.*

On the top you see the number of satellites that are visible above the elevation mask for a 24-hour session on the day you specified. The bottom shows the corresponding Position Dilution of Precision (PDOP), based on the number of satellites and their geometry.

{__15} *From this window you can probably get a pretty good idea of when, and when not, to collect fixes. You would avoid times of low satellite numbers and high PDOPs. (Leave this and subsequent windows open.)*

{__16} *From "Options" set the "SV Sample Rate" to every 10 minutes.*

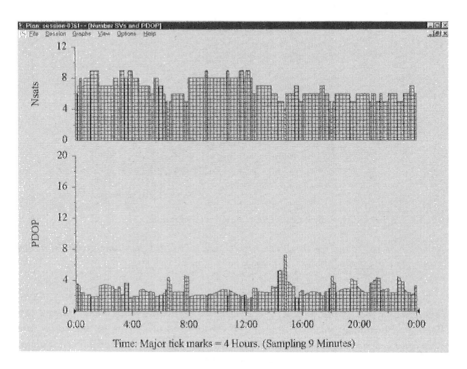

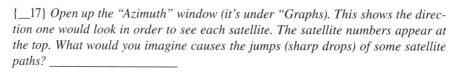

FIGURE 7.3 Graph of PDOP and number of satellites.

{__17} *Open up the "Azimuth" window (it's under "Graphs). This shows the direction one would look in order to see each satellite. The satellite numbers appear at the top. What would you imagine causes the jumps (sharp drops) of some satellite paths?* _____

{__18} *Open up the "Elevation" graph. You could use this information to pick a time when a cluster of satellites was high in the sky. Change the time period by picking "Mag" (for magnify) from the "View" drop-down menu. Magnify the time scale again.*

{__19} *Open up the "Satellites" graph. This shows the times at which each satellite rises and sets above the local horizon.*

{__20} *Pick a point on the time scale toward the right end, noting the time you choose. Click the left mouse button. Type the character "p." Note that the time scale shifts so that the time you picked is toward the center of the time scale. Now type "m." Now type "d" repeatedly. Note that you can get the day you indicated and all the next day.*

{__21} *You can also adjust the time by sliding the little arrows that bracket the time line. Try this. Use View ~ Redraw (or just type "r") to get back to the 24-hour time line. Now look at the drop-down View menu to see what items you have been selecting with "m," "d," "p," and "r."*

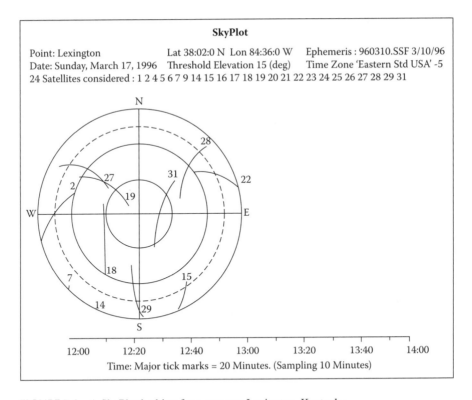

SkyPlot

Point: Lexington   Lat 38:02:0 N Lon 84:36:0 W   Ephemeris : 960310.SSF 3/10/96
Date: Sunday, March 17, 1996   Threshold Elevation 15 (deg)   Time Zone 'Eastern Std USA' -5
24 Satellites considered : 1 2 4 5 6 7 9 14 15 16 17 18 19 20 21 22 23 24 25 26 27 28 29 31

Time: Major tick marks = 20 Minutes. (Sampling 10 Minutes)

FIGURE 7.4   A SkyPlot looking from space at Lexington, Kentucky.

{__22} *Open up "SkyPlot," again under "Graphs." You see the track of each satellite in the time period shown at the bottom.*

{__23} *Look at the SkyPlot of Figure 7.4. It covers the same two-hour period (the two hours after noon on St. Patrick's Day of 1996) as the SkyPlot of the first figure in the book, Figure 1.1. But it looks different, in terms of where the satellites are. Why? _____.*

{__24} *Under "Graphs" choose "Tile 2 Column." If you haven't closed any windows you should see all the graphs you have opened on the screen, with identical time periods. Close some so that you only have four showing.*

{__25} *Choose "Redraw" under "View." This will restore the 24-hour time scale. And things will look pretty cluttered.*

{__26} *Click the title bar in any window except the active one. It will become the active window, and will fill the screen.*

{__27} *Under "View" pick "Close." Under "Graphs" pick "Close All."*

Looking at Tables in Quick Plan

{__28} *Choose "Report Type" under "Options." You want the report that shows changes in constellations.*

{__29} *Again under "Options," show the report. You could use the "File" menu to print it if you wanted to. Upon examination, you can see that this would be a pretty effective planning tool, complementing the graphs. Dismiss the report.*

{__30} *Just for illustration, change the "Elevation Mask" (under Options) to 45 degrees. Examine the same report and note the few times you would have enough satellites to collect data. After examining the table, choose the graph that shows the number of satellites and the resulting PDOP. Apparently those satellites between 15 and 45 degrees are vital to position finding.*

Experiment

{__31} *Make a SkyPlot. Note the broken red line between the two innermost circles. That is the 45-degree line. While all satellite tracks are shown, only the ones within that circle would be considered by a GPS receiver with the elevation mask set to 45 degrees.*

{__32} *Change the elevation mask back to 15 degrees. Close all graphs and tables.*

You can see the power of Quick Plan. It's also fun for those who like geography. Here are a couple more exercises.

{__33} *Under "Session" Pick "Create>>" to make a new point. Push Keyboard. Call the new point "North Pole" and make its coordinates 90°00'00' N and 0°0'0"W. Remove any existing points from the session and add North Pole. Set the elevation mask at 47°. Make a full screen SkyPlot. Note that the satellites never get above 47° in latitude. Note that there is an organized regularity to the paths of the satellites. Look at the Elevation graph. Bring up the Azimuth graph. Set the Elevation Mask back to 15°. Bring up the graph that shows number of satellites and PDOP. Tile 2 Column. Note that you can get good PDOP even at the North Pole. It truly is a* **Global** *Positioning System.*

{__34} *Edit the point of interest so you remove the North Pole and bring back Miami.*

When the Sky Isn't Clear

As you have no doubt discovered by now, when you are taking data with a GPS receiver, a major factor is sometimes obstructions that prevent signals from reaching the antenna. One of the really nice features of Quick Plan is that it lets you simulate, in terms of sky obstructions, the points at which you are going to take data.

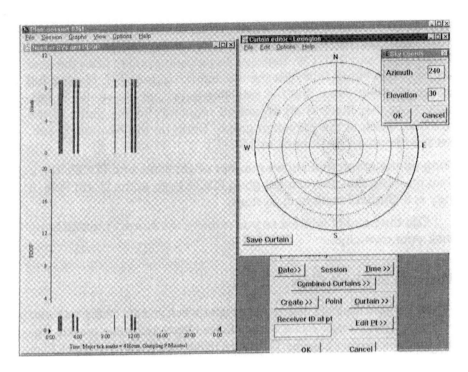

FIGURE 7.5   Simulating obstructions blocking the southern sky.

{___35} *Bring up the graph of Number SVs and PDOP. Tile it in 2 column format so it occupies half the screen. Edit the session (Session ~ Edit Session ~ Edit). Click "Curtain>>." Rearrange the windows so they look like Figure 7.5. In the Curtain editor click Options ~ Grid once.*

Suppose you want to represent an obstruction beginning at 120° azimuth that starts at 30° elevation, rises to 60° elevation in the due south, and returns to 30° elevation at a 240° azimuth.

{___36} *Position the cursor at Azimuth 120 and Elevation 30. (You can do this roughly from the diagram and check the Sky Coords box for the precise position.) Click. Now move the cursor (clockwise) to Azimuth 180 and Elevation 60. Click. Now to Azimuth 240 and Elevation 30. Click.* **Right** *click.*

{___37} *Look at the graph while you press Save Curtain. Note the dramatic change in the number of satellites and PDOP. Click OK in the lower right of the window.*

{___38} *Bring up a SkyPlot. The dark tracks indicate the positions of satellites blocked out by the curtain. Dismiss the SkyPlot.*

{___39} *In the Curtain editor window, click Edit ~ Clear. Answer OK. Then Save Curtain to save the cleared curtain which restores the obstruction-free situation so that no obstructions are saved with the point. Click OK under Sky Coords. Play*

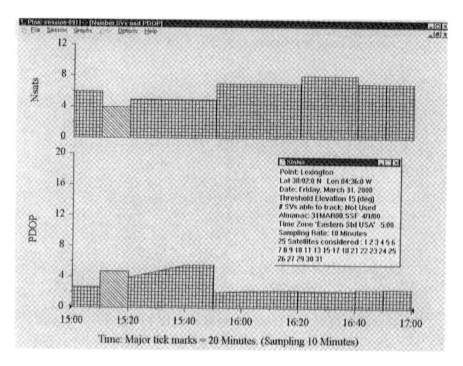

FIGURE 7.6   PDOP and number of satellites (Nsats) indicating when to and when not to take data.

some more, checking out the various options. If things don't seem to go right, exit the curtain editor and reinvoke it. At the end you need to clear the curtain and save the cleared condition. Otherwise the curtain stays associated with the point. Exit Quick Plan.

PROJECT 7B

This project will illustrate how taking data at the wrong time can result in loss of data. A route was driven at a time when only five satellites were visible. The same route was driven later when seven satellites could be tracked. The graph showing the PDOPs and number of satellites during the two-hour time frame that covered both data collection sessions is shown as Figure 7.6. You can note that at about 3:50 p.m. (1550 hours) the number of satellites is five and the PDOP is six. Just after that time the PDOP drops to approximately two and the number of satellites increases to seven. You would expect, in general, that a higher number of satellites would result in a lower PDOP, since a greater number of satellites allows for more choices of four-satellite constellations. In the following steps you will be able to see the effect of choosing a better time to take data.

{__1} *In Pathfinder Office, navigate to ___:\GPS2GIS\Planning and open the file From_UK_1. COR. Set up View ~ Layers ~ Features so that each Not In Feature fix is shown with a moderately thick red dot and that the dots are not joined. Set the coordinate system to be KY state plane north, etc. Set the time zone to Eastern Standard. Bring up the background files one.tif through six.tif. From the Position Properties window locate the First fix. Note that it was taken at about 3:30. When was the last fix taken? _____. Notice the sizable break in the GPS track near the southern end. More than half a mile seems to be missing just north of the highway interchange.*

{__2} *Open From_UK_2 (only) which constituted the same drive taken starting about 40 minutes later.* Now the trail of fixes is more complete. In fact, if you zoom up on the track where it appears to loop, you see that it shows an unexpected side trip that the car made: it turned into a bank driveway, circled the bank, turned right (heading north) then left on side roads to rejoin the main road going south. (The previous trip had done this as well but you couldn't see it because of the data loss.)*

{__3} *Look back at Figure 7.6. Starting the data collection session 40 minutes later made a real difference in the satellites available and the PDOP. And this made a significant difference in the data fixes that were collected!*

As with the rest of Pathfinder Office, Quick Plan has features we have not experimented with. Check out the rest of the menu options if you wish. Consult the Help files. Consult the manual.

Project 7C

In this project you will get set up with the current Trimble Planning Software, explore a Trimble GPS resources site, install an almanac from the Internet, and, if you wish, learn to use this software to provide the sort of information you looked at in Projects 7A and 7B.

{__1} *First check with your instructor or system administrator to see if the Trimble Planning software is already installed. It is likely to be in Programs > Trimble Office > Utilities > Planning. If not, you can download it from http://www.trimble. com/planningsoftware_ts.asp.*

{__2} *If necessary, install the software (perhaps with the help of your system administrator).*

---

* Okay, so it was a day and 40 minutes later. The same constellation was available, since the difference is only 4 minutes a day. The satellites arrive at the same point in the sky just 4 minutes earlier each day.

{__3} *From the same web site you will be able to download the latest almanac. Click on Trimble GPS Data Resources which will get you to http://www.trimble.com/ gpsdataresources.shtml. What resources can you get from this site?*

_____

_____

_____

_____

You will want the GPS/GLONASS almanac in Trimble Planning file format. Save it in ___:\GPS2GIS\Planning.

{__4} *Start the Trimble Planning Software. You will probably find it with Start > Programs > Trimble Office > Utilities > Planning.*

{__5} *Click on Almanac, then Load. Agree to clear any previous almanac. Navigate to ___:\GPS2GIS\Planning. Make the file name almanac.alm. How many total satellites of each type are in the current almanac?*

GPS _____

Glonass _____

Galileo _____

Compass _____ (Extra credit: What's Compass?) _____

WAAS _____

{__6} *Click on Help. If you want to use the Trimble Planning software, use the Help files to guide you to the sorts of information that you used in Project 7A. You could plan a future mission. Or, if it hasn't been too long since you took data in Chapter 2, you could look at the various planning parameters such as PDOP and number of satellites given by the planning software and compare them with what you recorded at the time.*

8 The Present and the Future

In which we examine several facets of GPS and GIS that are of current and future interest.

There was of course no way of knowing whether you were being watched at any given moment. How often, or on what system, the Thought Police plugged in on any individual wire was guesswork. It was even conceivable that they watched everybody all the time. But at any rate they could plug in your wire whenever they wanted to. You had to live—did live, from habit that became instinct—in the assumption that every sound you made was overheard, and, except in darkness, every movement scrutinized.

George Orwell, *Nineteen Eighty-Four,* **1949**

FIRST CONCERNS

The Orwell quote above, as it turns out, did not apply to 1984. But it might well apply to everybody in 2014. Or apply to you now, depending on who you are. Is privacy, and pretty much the bulk of the U.S. Bill of Rights, going to be lost? And with it our democracy? It's common to say, "If you haven't done anything wrong, you have nothing to fear if your movements and pronouncements are an open book." That might, only might, be true, if you trust the government. And thinking now of the Patriot Act in general, and the Valerie Plame incident in particular, I don't.

I'm hardly alone in feeling these concerns. Alan Cameron, editor-in-chief of *GPS World*, wrote an opinion piece (September 2008) titled "**Lost, Found, and in the Government's Pocket**" in which he, in part, said the following:

How do you feel about the government and its various enforcement branches knowing your whereabouts whenever they please? What about criminal or fringe elements, who often gain access to the same tools?

While GPS has made it harder to get lost, it also makes it harder to hide—or simply to remain as what used to be known as a private citizen.

As much as most of us—certainly most of us in the PNT industry—love technology, we cannot allow it to abrogate our constitutional protections. And yet this is about to happen—indeed, has already happened, by our default.

The skies are watching.

249

A corroborating view has been expressed by Lee Tien, senior staff attorney at the Electronic Frontier Foundation, in a quote from an NPR interview (again, courtesy of Alan Cameron and *GPS World*):

> There is a tendency these days to exploit whatever privacy invasion can be justified by a legitimate purpose such as safety or congestion or needing to pay a toll, and then, you've got this extra amount of surveillance, extra information, extra amount of invasion of privacy that's sort of comes with it—it's piggy-backed on it. And then, when it's sort of packaged in a good purpose like safety, people sort of let their guard down and start to accept those kinds of incursions.
>
> And the problem is that as you build these systems up and you do not design for privacy, then that amount of extra surveillance really leads you down to a slippery slope.

LOCATION-BASED SERVICES: ECONOMIC BOON?
INVASION OF PRIVACY? POTENTIALLY BOTH.

There is a new make-a-billion scheme on the horizon. It is called Location-Based Services (LBS). The bet is that your location at given times makes you eligible for getting services, or being made aware of services, that you will be willing to pay for, directly or indirectly. Look at some examples:

Your car quits (or hits a tree) in some remote area. Almost instantly people you would like to know about your misfortune can be made aware of it, and of where you are, within several feet. This system is already installed on many automobiles. It utilizes radio transmission of your position as determined by GPS.

You are navigating city streets toward a destination, according to aural instructions and a map display. Your in-car unit knows where you are by GPS and perhaps some additional gadgetry (like wheel counters and/or an inertial measurement system) that works when you go through tunnels and urban canyons. Except for the GPS, this system requires no inputs from outside. The GIS information comes from a "map" stored in the memory of the unit. But this capability could be combined with an LBS that warns of traffic congestion on your intended route and provides information about an alternative.

In the United States, your cell phone is required, by the FCC E-911 requirements, to be able to make your location known within 50 meters two-thirds of the time. So if you dial 911—whether you are outdoors, in a car, or inside a building—emergency personnel can find you. GPS plays a big part in this.

Your cell phone has been in standby mode—thereby transmitting its presence to a nearby cell tower. You suddenly hunger for a hamburger. You press a few buttons and a map is displayed on the screen of your phone, showing the locations of nearby eateries. Your handset is "location aware" so you can access such "location-based content."

You are in Japan in one of many cities. You need to see the location of an address and how to get there. You've paid a few dollars per month for a service that will do the equivalent of MapQuest or Google Maps, admittedly on a small screen.

> You are cruising down an interstate highway and a traffic situation develops that
> causes you to exceed the speed limit by 20 miles per. A week later a bill arrives
> for $146.50 for speeding—time, date, and place automatically recorded.

I don't think I need to belabor the point: The advantage, when you use GPS-enabled
wireless technology, is that "they know where you are." The disadvantage, and
maybe danger, when you use GPS-enabled wireless technology, is that "they know
where you are."*

ON THE HORIZON: TRENDS

Some time ago, in a GPS trade publication, I read something to the effect that GPS
receivers are becoming so integrated with other equipment that GPS itself is becoming virtually invisible. "Oh, great!" I thought. "Just when I've spent all this time and
effort learning about it." You may have a similar reaction.

Upon reflection, I realized that the observation, while true for many uses of
GPS, probably will not apply to GIS applications. Most users of GPS make immediate use of the satellite signals and do not record them for future processing. The
use GIS professionals make of GPS is for the development of data sets, whose
quality must be high and which will endure for months, years, or even decades.
But it is certainly useful to look at the trends in GPS use across the board, to more
fully understand this amazing phenomenon. The following seem to me to be fairly
safe predictions. (Since this is being written in March 2009, at either the height (or
the start) of a world financial crisis, I'll order this list in terms of those elements of
GPS that might aid the people of the world by helping make better use of Earth's
resources.)

1. Air navigation will be radically transformed.
2. Ground vehicle navigation will result in more efficient use of resources.
3. "Precision Farming" will play an increasingly important part in maintaining and enhancing the world's food supply.
4. Reliability will increase dramatically for some applications.
5. Civilian and military sectors will increase cooperation; the civilian side of
   GPS becoming relatively more and more important.
6. New equipment configurations for database building, database maintenance, and navigation will be developed as technologies merge.
7. GPS will be combined with other systems to provide positional information
   in places of poor or limited GPS reception.
8. Emphasis will shift from using GPS to provide information to humans who
   control processes and artifacts, to direct control of those processes and artifacts by GPS.
9. GPS will be the principal mechanism for distribution of accurate information about time.

---

* The Web site, if it still exists, http://www.themobiletracker.com/english/index.html, is a joke. But for
  how long?

10. "Monumentation"—the activity of providing markers on the Earth to indicate exact geographic location for reference by those who need such information—will increase and change radically in approach.
11. Accuracy will increase, both gradually and by significant steps. The time required to obtain a fix of a given accuracy will decrease.
12. The U.S. NAVSTAR GPS will be modernized.
13. More applications of GPS will emerge; those in use will become more extensive.
14. There will be competition for the radio spectrum frequencies that GPS uses; if the spectrum is shared there is the danger that the quality and reliability of GPS signals could be affected.
15. New satellite navigation systems will emerge, perhaps complementing NAVSTAR, perhaps competing with it.

A brief discussion of some of these trends follows.

AIR NAVIGATION WILL BE RADICALLY TRANSFORMED

The Wide Area Augmentation System (WAAS) based on GPS is now a reality. Today, to keep track of a flight, a flight controller's radar sends a pulse which strikes the airplane and triggers a "transponding" pulse from a radio carried in the aircraft. This gives the controllers on the ground the 2D information they need to tell the pilot where to fly to avoid other aircraft. But the altitude information has to come from the airplane's own altimeter, whose readings are based on barometric pressure. With GPS and WAAS, the airplane's electronics will simply know where it is, in all three dimensions, and that information will be sent directly to the controller. This will come to pass over time as aircraft electronics are upgraded.

Already the VHF OmniRange (VOR) is being used less and less, as aircraft navigate based on GPS. One concept that is being discussed is doing away with many of the airways. At present, aircraft operating under instrument flight rules (IFR, where pilots whose planes are in the clouds avoid hitting the ground by reference to charts and instruments, and avoid hitting other planes by being told where those planes are by a radar controller on the ground) follow highways in the sky. And just like highways on the ground, they do not usually lead directly from the trip's origin to its desired destination. One reason for the detours is collision avoidance. A second is that navigating the airplane is based on having to go through intersections (where the limited-range, ground-based antennas are). The combination of GPS and "real-time GIS," which shows the location of other aircraft, can obviate both of these factors.

If every plane could fly "direct," detouring only when the GPS/GIS system suggested that the current course might result in bent metal, vast amounts of time and resources would be saved.

In addition to navigation en route, GPS systems, augmented by real-time differential techniques, have shown abilities to provide positional information of less than half a meter to landing aircraft.

Vehicle and Marine Navigation Will Be Improved

Something on the order of several million in-car GPS navigation systems are being installed each year. Not only do these systems allow navigation based on the static facts of where the roads go but also on the dynamic facts relating to where the other traffic is. The city of Tokyo benefits from such a system, as do some U.S. cities.

In addition to vehicle route guidance, intelligent transportation systems will provide other benefits, such as fleet vehicle dispatch and tracking, and emergency notification of police and tow trucks. Before you cheer too loudly, however, realize that without proper safeguards, Big Brother may know where you are, where you are going, and how fast. Already, automated systems exist in the United States that scan license plates of speeding cars and send a ticket to the owner.

Precision Farming

Yours truly had the amazing and scary experience of flying in a helicopter during a simulated fertilizer spraying run. (As a "fixed wing" pilot I regard any flying condition in which the ground is less than 50 feet away, except when landing or taking off, with extreme trepidation.) The helicopter was equipped with the Trimble Trimflight GPS system that allowed the pilot to fly precisely along a predetermined grid superimposed on an image of the target field, shown to him by a computer display. Further, a set of lights told him if he was left of, right of, or on course. Normally, the navigation involved in spraying a field is accomplished by (1) the pilot remembering approximately where he was on the last pass (in addition to flying the bird and keeping it out of the wires at the end of the field), and (2) some poor fellow on the ground, covered with crud, waving a white flag.

Not only was the spraying operation made safer by Trimflight, it was also made more effective. Earlier, during harvesting, a combine had collected location-specific yield data on the field, showing areas of more and less fertility for the crop. Chemical soil samples were taken, indicating site-specific deficiencies. So when the spraying operation took place, the amount of product could be metered depending on the position of the aircraft. The results: more stuff in the right spots, less pollution, less expense.

Less dramatically, the same sort of application of pesticide or fertilizer can be done by ground-based equipment; that has the advantage of not having the "**LOW FUEL**" light flashing, as we did on our flight.

All in all, precision agriculture is one of, if not the, most significant GPS applications, whether done with aircraft or, more usually, with ground-based equipment.

System Integrity and Reliability—Great Improvements

While you might think that the maximum safeguards would be provided for a system designed to aid a warrior with a mortar or a nuclear-tipped ICBM, the fact is that GPS position data can suddenly disappear for periods of a few seconds to a few minutes. That state of affairs is not going to be allowed for some important civilian applications. You will understand the need for reliability immediately if you imagine yourself as a passenger in a landing commercial aircraft. The plane is 50 feet above a runway in a blinding snowstorm. That's no time for GPS to take a breather.

It is going to be possible for ships to navigate in shallow water by knowing their vertical height and having an integrated survey of the bottom of the water body. On-the-fly marine navigation requires 3D GPS decimeter (that's 10-centimeter or 4-inch) accuracy. Again, everything has to work right and keep working.

GPS will be vital to all sorts of civilian and commercial endeavors, so the systems built around it simply must be highly accurate and reliable.

CIVILIAN AND MILITARY INTERESTS WILL CONTINUE TO COOPERATE

Presidential Decision Directive 63, issued March 29, 1996, declared that "a permanent interagency GPS Executive Board, jointly chaired by the Departments of Defense and Transportation, will manage GPS and U.S. Government augmentations."

In September 2000 the Civil GPS Service Interface Committee announced that DoD and DoT agreed to seek greater involvement from civilian federal agencies in the day-to-day management of GPS. Civil representation is to be sought throughout all stages of the DoD acquisition, requirements development, and planning, programming, and developing process.

Once the civilians and the military agree on a few more issues related to GPS accuracy, coverage, and integrity, GPS will be the primary U.S. government radionavigation system well into the next century. The satellites which will be launched to provide GPS information in 2015 and beyond are already being designed. Swords are being beaten into plowshares, but the changes will not come rapidly, for both political and security reasons.

A real breakthrough would be the development of a way to jam GPS signals in a given geographic area, thus denying an enemy any use of the civilian GPS signal.

MORE EQUIPMENT AND SOFTWARE CONFIGURATIONS WITH GPS AND GIS

The direct integration of GPS and portable computers running GIS has been a reality for some time, allowing new facility in navigation and data collection. The company GeoResearch developed GeoLink software which linked GPS input to a microcomputer running GIS software.

As you may recall from the flying tour example of Chapter 7, Trimble Navigation made a software product called Aspen that runs on a portable computer. A PCMCIA GPS receiver card (also called Aspen) is inserted into a socket in a laptop, notebook, or "pen" computer and an antenna is attached. When the Aspen software is running it looks a lot like Pathfinder Office with some additional menus for controlling the GPS receiver and recording data.

A sister product, "Direct GPS," integrates GPS signals, RDGPS correction broadcasts, and turns a standard computer into a datalogger and navigator. Direct GPS displays position information, with ArcView providing the software context.

Another configuration gaining in popularity is the use of a handheld touch screen computer running ArcPad software from ESRI. Basically, you can download ESRI or other data files and image files to the handheld computer, attach a GPS receiver to it, and go into the field. GPS shows you where you are in the context of the area of interest. Once there you can collect data, correct data, or perform other database maintenance functions.

If your requirements are more in the navigation line, you may obtain a GPS receiver, software, and map data from DeLorme and hook it to a Personal Digital Assistant (such as a Palm Pilot).

A truly intriguing method of data collection and database maintenance is available from Datria Systems. It combines GPS capability with voice recognition software. In use in several metropolitan areas, it allows a person to collect data by simply speaking into a microphone while walking or driving with the system. The data collector might say, "stop sign down, 4 feet right" and later "broken pavement 20 feet ahead" and this information would be recorded in the unit. Back at the office the synchronized voice data and GPS fixes are uploaded into a PC that turns the voice print into written text and writes and routes job orders to the departments responsible for signage and pavement repair. It is easy to believe that this system lives up to the claim of making such data collection 10 times faster than even "traditional" GPS data collection techniques.

As previously mentioned, ESRI has produced an extension called the Tracking Analyst that allows real-time tracking of GPS receivers, or "playing back" data taken by such receivers. Basically, the Tracking Analyst allows the user to plot GPS data as it is being acquired, by virtue of radio transmissions, perhaps by communication satellite or cellular telephone, of the GPS data to a location that then relays these data to the software. Information from the Tracking Analyst can be routed to the ESRI Internet Map Server (IMS) to make map data available on the Web. Some interesting projects involving wildlife tracking (elephants and elk) are underway using this technology. The Magellan receiver featured in this text contains a camera. And Kodak makes a digital camera to which you may attach a Garman GPS receiver. This can create a watermark on the image itself, or become a part of the attribute data in the image file. Other camera manufacturers will make such products and interface them with other GPS units.

Sometimes you want the GPS position of an object that is some distance away and not convenient to get to or, perhaps, inaccessible. Maybe you want the x, y, and z of Thomas Jefferson's nose at Mount Rushmore, in the Black Hills of South Dakota, without risking life and limb to record the data. Some GPS receivers may be outfitted with a laser range finder that can report the distance and direction of an object at which it is aimed. An accuracy of half a foot over a distance of 2000 feet can be obtained.

There's a gadget that determines the character of light that goes through the leaf of a plant—thus revealing information about the chemical composition of that leaf. Hooked to a GPS receiver it allows the location of the plant to automatically become part of the data file.

Want a small GPS receiver? Furuno USA makes one that is 28 mm × 21 mm × 10 mm and weighs 18 grams. That's a little more than an inch by a little less than an inch by less than half an inch thick—and four-hundredths of a pound.

GPS Combined with Other Systems

It has been commented that GPS bears a resemblance to sanitation plumbing of a hundred years ago: basically an outdoor activity. But often we want to know where we are when we don't have the luxury of a line-of-sight relationship with four or more satellites. In tunnels or mines or caves, for some examples—even in deep valleys or on urban streets—GPS operates at a disadvantage. One way to locate your position, assuming you knew it at some time in the past, is an Inertial Navigation System (INS). An "INS" is a mechanical apparatus, whose heart is a set of spinning gyroscopes; it can detect very small changes in its position and report these to an operator. Marine navigators can place position data into such systems while still in harbor, sail for three months, and return to the point of origin, with the "INS" indicating the original settings within a hundred yards or so. Certainly it is possible for a vehicle which is mapping a road system to have an INS on board to see it through tunnels, or valleys between skyscrapers.

The Ohio State University Center for Mapping has developed a van for delineating highways that combines a number of position locating techniques and systems with photographic or videographic information, showing pavement condition, signs, intersections, and the like. The information can be installed in a GIS. Later, a point on an interactive computer map can be selected and the user can "drive" along the road, seeing the features.

Monuments Will Be Different, and in Different Places

Things are changing at the National Geodetic Survey agency office—more in the past ten years than in the preceding two centuries (Thomas Jefferson set up the Survey in 1807). Monuments used to be on the tops of mountains, since line of sight was vital to the surveying profession. Line of sight is still vital, and still in the upward direction, but toward the satellites. And instead of static monuments we can expect radio-dynamic ones, that announce where they are. Further, the physical monuments will be more accessible and more closely spaced—and more accurate, having been "surveyed-in" with GPS techniques.

The charted positions of many places have been changed based on GPS information. Numerous islands have been moved from their originally charted positions. The coordinates of the city of Hannover, in Germany, shifted a couple hundred meters to the south as worldwide systems replaced national ones.

GPS Will Become the Primary Way to Disseminate Time Information

A GPS receiver is one of the most accurate clocks in the world, if it has continual access to the satellites. The forte of a clock in a GPS receiver is short-term accuracy, not long-term consistency. People who really want to know exactly what time it is can set up a base station over a known point and analyze GPS signals for time instead of position. GPS time does differ from UTC time by an integer number of seconds, less than 20 for the foreseeable future. There are many applications for which coordinated time is vital. Examples: making the telephone system, the banking system, and the Internet run smoothly. The interior core of Earth can be probed because it is possible to track earthquake shock waves from one side of the planet to the other—thanks to being able to know the exact time in each place. A downside: It is feared that GPS time will become so ubiquitous that the expertise for the building of atomic clocks will disappear, as the demand for those shrinks to only a few a year.

BETTER ACCURACY

In addition to the gradual increases in accuracy that will be brought about by improved equipment and processing methods, several other factors will considerably improve what we may expect of GPS in the future. As you know, military receivers are able to be more precise than civilian ones for a variety of reasons. One is that they receive on two frequencies. Comparison of the two signals allows a better estimation of the error caused by the ionosphere. Having two frequencies will provide civilian receivers with the ability to immediately compensate for one of the larger sources of error: ionospheric delay. As some point new satellites will broadcast an additional signal (named L5), dedicated to civilian use, on 1176.45 MHz. The signal may be four times stronger than the existing civilian signal (L1); L5 will be more resistant to interference. While the primary function for the L5 signal is "safety of life" for aircraft navigation, it will provide a robust additional signal for all users.

By having L1, L2, and L5 signals, the stand-alone (nondifferential) real-time GPS user will have:

- Signal redundancy
- Better position accuracy
- Improved signal availability
- Warning when the GPS signal might become unhealthy (called signal integrity)
- Improved resistance to interference from other radio signals
- Fewer or no breaks in service

Differential GPS will improve as well over the years, but not as dramatically. And no one has suggested yet that to use GPS for GIS you can do without it.

The wide availability of DGPS signals will increase and improve, as will the types of origins of such signals. Between the Coast Guard Beacon network, commercial firms offering FM sub-carrier signals, and the DGPS signals that come via communication satellite, DGPS is pretty much available in the United States. For really precise work, DGPS is required anyway.

FASTER FIXES

While GPS at present can provide surveyors with the centimeter-level accuracy they require, the length of time they need to "occupy" a location is usually prohibitive for routine land surveying activity. (Employees of surveying firms sometime comment that the term GPS means "Get Paid for Sitting.") The kind of accuracy surveyors need requires that the receiver stay in constant touch with the satellites, so that they may actually count the number of wave cycles of the radio signal between the antennas. This is termed "carrier-based" position finding, as differentiated from the GPS work (code-based) that you have studied. In areas or conditions where the receiver may "lose lock" on the satellite, the process can be not only long, but frustrating. Better equipment and the availability of carrier phase DGPS signals will make GPS viable for many more surveying applications.

GPS manufacturers are aware that the speed with which a receiver can acquire enough satellites to know where it is, bears on marketing as well as position finding. If they do well in this arena they advertise their "time to first fix."

NAVSTAR GPS Modernization

You might think of GPS as a modern phenomenon, but there have been GPS satellites in orbit since 1978. The U.S. Congress, known more for its political intrigues and infighting rather than its concern for the citizenry, debates the issue of whether to adequately modernize NAVSTAR. Probably it will happen because GPS is pretty much a military necessity.

The (1) uncertainty of whether the United States was going to guarantee sufficient civilian control over NAVSTAR to provide a reliable civilian signal, and (2) the questions of modernization concerned (irked) those in other countries—particularly Europe, where a move is underway to build a European GPS called Galileo.

What's in the offing, if all goes well, is a $1 billion–plus modernization program that is badly needed if GPS is to serve the coming two or three decades as well as it has served the last one. For civilian use of GPS, the modernization program will look something like this:

2009–2011: Satellites launched in this period broadcast the second civilian signal (L2), as well as L1. In April 2009 the first satellite to broadcast the "safety of life" signal (L5) went into operation.

2012–future: GPS III satellites. More power plus other enhancements conceived of in the next decade.

The second civil signal is a big, big deal. It is at 1227.6 MHz, called GPS L2. The primary civil signal, on L1, is at 1575.42 MHz. The speed of light and radio waves in a vacuum is the well-known constant "c." But when electromagnetic radiation travels through substances (e.g., water, glass, air) the speed changes—and the amount of change is dependent on the frequency. Having two signals of different frequencies to work with allows a properly programmed GPS receiver to "calculate out" most of the error caused by the atmosphere. The atmosphere (the ionosphere, actually) causes most of the error that there is. If the average error from all sources, considering only one frequency, is, say, 11.5 meters it could be reduced to about 4.7 meters—quite an improvement! As of September 2008 six GPS satellites were broadcasting the L2 signal.*

The third civil signal, to be broadcast at 1176.45 MHz (GPS L5), was also announced in 1998. It has a somewhat different purpose. The signal at L2 will provide considerably increased accuracy, but is subject to interference from certain radar transmitters. L5 does not have this problem. The signal at this frequency will be designed for safety-of-life services.

But GPS modernization has proceeded in fits and starts and will continue to do so—dependent as it is on U.S. political institutions. For example, the Congress stopped

---

* Specifically, PRN numbers 7, 12, 15, 17, 29, and 31—if your planning is that detailed. Check http://en.wikipedia.org/wiki/List_of_GPS_satellite_launches for more up-to-date information.

funding Nationwide Differential GPS (NDGPS) in 2007 despite its growth and success in the past 10 years. (NDGPS is also referred to as the Coast Guard differential system.) The system could provide dual coverage, high-precision augmentation of the GPS signal for aviation and maritime applications across the United States.

APPLICATIONS: NEW AND CONTINUING

New applications appear every day, from aiding farmers in feeding the hungry people of the world by enhanced agricultural techniques to golf cart tracking and pin ranging. Animals can be tracked by researchers. Polar ice sheets can be monitored. Blind people can be guided. Rescue helicopters can be directed. Solar eclipse tracks and weather information can aid navigating astronomers. Chase cars can follow hot air balloons, even when they can't be seen. And in space: some satellites carry GPS receivers! What better way to report on exactly where they are?

The magazine *GPS World* reports monthly on new applications, and sponsors an applications contest once a year. And, of course, the World Wide Web has information of interest to those searching for information on GPS.

COMPETITION FOR THE RADIO SPECTRUM

If your campus is like mine, every third pedestrian student has a cell phone glued to his or her ear. If your town is like mine you would be wise as you drive to keep a wary eye out for vehicles operated by people steering with one hand and chatting wirelessly, perhaps with their children or brokers. To find a place to channel all that telephone traffic, the people who make money selling the service have to find "bandwidth" somewhere. Guess what one of their targets is. Right. The GPS frequencies.

Toward the end of the 1990s, application was made by the mobile satellite system (MSS) industry to share the L1 band for downlinks of voice and data. The application almost succeeded, despite the dangers that such frequency sharing might have caused possible interference with the civil GPS signal. It has apparently been turned back by the World Radio Conference. The threat seems to have diminished.

Another threat to interference-free GPS signals may come from ultra wide-band (UWB) electronic devices that may undermine GPS reliability. These devices may be used for the precise measurement of distances, for obtaining images of objects buried underground, or present within buildings; their bandwidths exceed 1 gigaHertz, but they use that spectrum in very specific ways that may or may not cause interference with other signals. The argument is that they will coexist with GPS because "they create no more radio interference than a hair drier" was attacked by Charles Trimble, longtime and well-noted GPS expert who identified five crucial differences between a UWB device and a hair drier:

- A hair drier is not usually on all the time
- It's rarely portable
- It has no antenna
- It is not connected in networks
- You can't convert a hair drier into a GPS jamming device

OTHER COUNTRIES, OTHER SYSTEMS

Because the United States has been very proprietary with the more accurate aspects of the NAVSTAR GPS, people in other parts of the world look with concern at hanging their navigation, timing, and position recording capability on a system over which they have no control. For some time now, the European Union has been developing its own Global Navigation Satellite System (GNSS) called Galileo. Why would another "GPS" be necessary, or even useful? Besides the issue that other countries do not want to depend on the United States and its fickle political institutions for quality position information in the coming decades, NAVSTAR GPS has a number of other shortcomings that make development of another system useful.

Because it will be designed from the ground up as a utility, Galileo may

- Allow more accuracy
- Be more reliable
- Admit to the user immediately when it fails
- Be more secure

Two Galileo satellites have been launched. It appears that Galileo will be 'interoperable' with NAVSTAR. With the two systems working together we could easily see a 50-plus satellite constellation that would provide marvelous availability, accuracy, redundancy, and integrity.

Today a number of events suggest that the direction of the future of GPS (or to use the more general term GNSS (Global Navigation Satellite System—they still haven't got "timing" in the name)) is not completely predictable. Although Galileo seems still up in the air, the European Space Agency does operate the SBAS European Geostationary Navigation Overlay Service (EGNOS).

Of course, there is already another GPS system up and running: GLONASS has had almost the same number of operating satellites as NAVSTAR, and has been around about as long. (The first GLONASS satellite was launched in October 1982.) But the Russian program has suffered greatly in the years 2000 to 2007 due to financial difficulties. During the year 2000 the system fell to only eight operating and healthy satellites. In April 2009 there were 24 GLONASS satellites operating so the system was close to full operational capability. Receivers are available that use both GPS and GLONASS satellites.

China surprised everybody when it launched a test navigation satellite named Beidou—their name for the Big Dipper. A second Compass satellite was launched in April 2009. The Chinese seem intent on their own system: Compass. There is considerable concern about Compass, because of compatibility issues with NAVSTAR and Galileo. Japan, on the other hand, seems content to make use of the GNSS that other countries are planning or maintaining, but to augment these systems with an SBAS. The Japanese are installing 1.5 million car navigation systems a year.

Japan's interest has been sharpened by the number of large earthquakes and other disasters they have experienced. Can GPS aid in predicting earthquakes? They certainly think so. While they are not intending to add to the number of GNSSs, they do have plans for use of satellite positioning systems in aircraft and automotive transportation: the Multi-function Transport Satellite Service (MTSAT), which is a geostationary satellite overlay service.

GPS: Information Provider or Controller?

There will be an increasing number of applications in which GPS signals control equipment directly, rather than going through a human "middleman." In such joint GPS/GIS uses as fertilizer or pesticide application, the automated system may steer the tractor while the farmer rides along simply for reasons of safety. Carving out roadways or laying pavement may be conducted in similar fashion. The Center for Mapping at Ohio State University boasts a system that can put a bulldozer blade in the correct position with an accuracy approaching 1 centimeter. Land fill compaction can be accomplished by running over the trash with a heavy tractor, guided by GPS, with an operator on board only to make sure nothing goes wrong. Sufficient precision for such activities may be assured by GPS signals from pseudo-satellites on the ground or SBAS.

THE FUTURE—WHO KNOWS?

If you are using this text after 2009 you may well know more about what is happening then than is printed here. And what is said here may be wrong or completely out of date. That's the way of accelerating technologies like GPS.

Appendix A

Sources of Additional Information about GPS

SOME PLACES TO LOOK FOR ADDITIONAL GENERAL INFORMATION ABOUT GPS AND ITS APPLICATIONS

GPS World, monthly journal, Advanstar Communications, http://www.gpsworld. com

Integrating GIS and the Global Positioning System, book, Karen Steed-Terry, ESRI Press, http://www.esri.com

GPS for Land Surveyors, 3rd edition, Jan Van Sickle, CRC Press, http://www. crcpress.com

National Executive Committee for Space-Based Positioning, Navigation, and Timing, pnt.gov

If you are interested in more Internet sites, you can go to a Web search engine (e.g., http://www.google.com) and type *Global Positioning System* into the search field. That will give you the possibility of about 400,000 pages to look at, which ought to use up the better part of an evening.

FOR TECHNICAL INFORMATION ON GPS ORBITS, TIMING, SATELLITE CONDITION, AND SO ON

Canadian Space Geodesy Forum, gauss.gge.unb.ca/CANSPACE.html
U.S. National Geodetic Survey (NGS), http://www.ngs.noaa.gov/orbits/
Scripps Orbit and Permanent Array Center (SOPAC), csrc.ucsd.edu
U.S. Naval Observatory (USNO), tycho.usno.navy.mil/gps.html

FOR INFORMATION ABOUT GLONASS

http://en.wikipedia.org/wiki/GLONASS

FOR INFORMATION ABOUT GALILEO

Galileo's World, quarterly magazine, Advanstar Communications, http://www.galileosworld.com

Galileo Program Office, http://www.galileo-pgm.org

GLOSSARY OF GPS-RELATED TERMS

Search the Internet for "GPS glossary."

Appendix B

*Form for Exercises, Lab Practicals, and Exams
Using Data Sets on the CD-ROM*

For the assigned folders or files, please provide the following information:

File of Folder designation(s): _____ Receiver Type: _____

File Date (dd/mm/yy): _____ Duration of file(s): _____

Number of fixes _____

Corrected file? Yes (__) No (__)

Time of first fix: _____ GMT _____

Local time of last fix: _____ Local

General location of data collection: _____

Specific site of interest, if any: _____

Positions of first fix and last fix:

(WGS84: latitude, longitude, altitude [meters]) _____, _____, _____

(WGS84: latitude, longitude, altitude [meters]) _____, _____, _____

The receiver antenna was: STATIONARY (__) MOVING (__)

If the antenna was STATIONARY, give these statistics:

Average northing, easting, altitude: _____, _____, _____

Standard deviations of the averages: _____, _____, _____

If MOVING: Largest distance between fixes (specify units): _____

Probable method of antenna conveyance: _____

Index